90 0614425 6

KV-545-671

WINGS ACROSS EUROPE

WITHDRAWN
FROM
UNIVERSITY OF PLYMOUTH
LIBRARY SERVICES

Wings Across Europe

Towards an Efficient European Air Transport System

University of Plymouth
Library

Item No.

KENNETH BUTTON

ASHGATE

© Kenneth Button 2004

All rights reserved. No part of this publication may be reproduced, stored in a retrieval system, or transmitted in any form or by any means, electronic, mechanical, photocopying, recording or otherwise without the prior permission of the publisher.

Kenneth Button has asserted his right under the Copyright, Designs and Patents Act, 1988, to be identified as author of this work.

Published by
Ashgate Publishing Limited
Gower House
Croft Road
Aldershot
Hampshire GU11 3HR
England

Ashgate Publishing Company
Suite 420
101 Cherry Street
Burlington, VT 05401-4405
USA

Ashgate website: http://www.ashgate.com

University of Plymouth
Library

Item No.
9006144256

Shelfmark
387. 7094 BUT

British Library Cataloguing in Publication Data
Button, Kenneth John
    Wings across Europe : towards an efficient European air
    transport system
    1.Airlines - Economic aspects - Europe 2.Aeronautics,
    Commercial - Economic aspects - Europe
    I.Title
    387.7'094

Library of Congress Cataloging-in-Publication Data
Button, Kenneth John.
    Wings across Europe : towards an efficient European air transport system / by
Kenneth Button.
        p. cm.
    Includes bibliographical references and index.
    ISBN 0-7546-4321-2
  1. Airlines--Europe. 2. Aeronautics, Commercial--Europe. I. Title.

    HE9842.A4.B89 2004
    387.7'094--dc22

                                                                    2004012701

ISBN 0 7546 4321 2

Printed and bound in Great Britain by TJ International Ltd, Padstow, Cornwall

Contents

List of Figures

List of Tables

Preface

This book grew directly from a report produced for the Association of European Airlines in 2002. The terms of reference for this report were very broad and basically amounted to looking at the longer-term future of European air transport.

This was rather adventurous for the Association in two ways. First, the remit was directed at longer-term problems at a time when the airline industry was suffering from the combined effects of a cyclical down turn in economic activity combined with the aftermath of the September 11th attacks on the US. Most industrial associations content themselves with fire fighting rather than considering broader issues that confront their members. Second, there was no effort to impose ideology or to force the work towards a lobbying position. Indeed, many European airlines may have been disappointed that the report did not look in more depth at internal matters of airline management. But in fact most industries contain a mixture of well and badly managed companies (and perhaps more importantly, lucky and unlucky companies) and airlines do not seem to defy that rule. More important in the longer-run is the institutional structure in which the companies offer their services.

The contents of this volume are thus a slightly up-dated version of the AEA report with some restructuring material to transform it into a book format. I would like to thank in particular Roger Stough and Kingsley Haynes who were involved in the preparation of the initial report, and Julianne Lammersen-Baum, who did much of the final editorial work on the book manuscript.

Chapter 1

Introduction

Aims of the book

Air transport is a major industry in its own right. But globally it is also the fastest growing mode of transport for both passengers and freight. As such it represents a large employer of labor, is at the forefront of many technological developments, and indeed is often a pioneer in adopting such innovations. It is the source of important economic stimuli for local and regional economic development and it often provides essential long distance access for more peripheral areas, serving as a key lubricant to the economic system.

At the same time it is a maturing industry with established technology and an accepted institutional structure. Further, it is by most measures the safest way to travel and there are continual technological and managerial developments to reduce any adverse environmental effects. In summary it is a part of modern society.

Air transport has also been the subject of major economic regulatory reforms over recent times. After many decades of fairly strict economic regulation there has been a gradual liberalization process. These changes, however, have not been consistent across markets. While the 1978 US Airline Deregulation Act was at the forefront of liberalization of the US economy, changes elsewhere have been less dynamic and often less complete.

The European Union (EU) was very slow to embody air transport in its Common Transport Policy (CTP). This reflected, in part, the still relatively small role it played in economic interactions in Europe in the formative years of the EU, the 1950s and 1960s. That situation changed significantly in the late 1980s and 1990s when national policies for air transport were gradually brought within the broader framework of EU policy. The need for a coordinated infrastructure policy, especially regarding air traffic control, furthered the role of the EU. This process is an on-going one as matters of authority over external EU policy continue to be debated.

This study seeks to examine the current state of European airlines – mainly, but not exclusively, those within the EU and the European Economic Area (EEA). It seeks in particular to determine if the current institutional structure provides a sustainable basis for the continued vitality of air transport as a facilitator of economic development and if it can serve as an input into wider matters involving the social and political integration within an increasingly politically and economically integrated Europe.

The book does not seek to offer any detailed forecasts for the future of European air transport, nor does it make any effort to identify 'winning' airlines in the commercial environment that now determines the world in which airlines live. External factors influencing the industry, coupled with its own internal dynamism, makes such exact predictions impossible. The complexity of the industry and the markets it serves add to the problem of being a seer. Rather the book focuses on the evolving industrial structure and the market for airline services.

The problems confronting airlines

The study comes at a time when many European airlines, like numerous carriers in other markets, have experienced, and some continue to experience, severely depressed markets with negative consequences for their financial performance. This is due to the combined results of regular business cycle effects and as a consequence of the more immediate repercussions of the attacks on the World Trade Center in New York and the Pentagon in Washington in September 2001. The SARS epidemic also had serious adverse effects in Asian markets in particular, and the Gulf conflict has had its own geographical effects. There are also emerging issues that are putting more general pressures on airlines, including increasing demands to fulfill societal environmental ambitions (Upham et al, 2004).

The analysis here, however, is primarily concerned with longer-term trends in European air transport, and in particular how these trends are affecting airlines. Whilst recognizing that periodic shocks to the economic system do occur, it focuses on the more conventional economic forces that affect the sector.

The world's air transport markets are still finding it difficult to recover from the events of September 2001 as a combination of factors have reduced the willingness of people to travel and a general slowing of the world's economy has constrained growth in demand for freight services.

In the short-term, a global picture showed that total scheduled air passenger traffic in June 2002 was 11.1% lower than in the previous year. In 2002, a net financial loss of $US12 billion was recorded by member airlines of the International Air Transport Association (IATA). More specifically for European carriers, passenger traffic was down by 7.8% within Europe and in important markets such as the North Atlantic it was down 18.6%. In terms of freight, while global carriage virtually returned to its June 2001 level within a year, it was 18.6% lower in Europe and 3.8% lower on the North Atlantic. Recovery has been slow and in 2002 scheduled traffic was down a further 13.8 million passengers.

There has been some small relative improvement since that time, although the spread of SARS, and the military action in Iraq, along with poorly performing European economies, introduced new challenges for the airlines. If nothing else the decline was less in 2003 with a further fall of 2.4% in global traffic.

The impacts on individual European airlines varied considerably, and this continues to be the case. The particular mix of markets served can explain some differences in the performance of specific companies – for example, having a strong North Atlantic presence has not always been helpful. But the airlines have themselves responded in different ways in their efforts to cut costs and to attract traffic. In particular, many carriers have significantly cut capacity. Unlike US carriers, very limited public funding was made available to European carriers after September 11th to compensate for the immediate impacts of closure of US markets, and there was no form of guarantee system to assist in overcoming cash flow problems in the intermediate term.

The outcome has been a severe financial crisis for most major European carriers. Overall scheduled European airlines lost US$3.02 billion in 2001 and US$0.87 billion in 2002. Sabena went bankrupt shortly after the crisis emerged, and other network carriers curtailed some of their operations, contracted, and began a major program of restructuring. There was an 8.7% capacity cut by the major European airlines in 2002.

The situation in 2003 saw little respite and although not all the figures are in at the time of writing it is clear that several major European carriers are taking time to recover: British Airways for example, made a profit of about US$1.0 billion, but Lufthansa experienced a US$1.2 billion loss mainly because of problems in its non-core airline catering and travel businesses, and Alitalia experienced a US$500 million operating loss in 2003.

These are, therefore, trying times for much of the European airline industry. But there is evidence that the current crisis may be a catalyst for bringing forward more fundamental structural problems in the industry. In particular, there have been institutional changes that have affected the way in which air transport services are provided in Europe and the ways in which the airlines interact with their suppliers and customers.

Some of these changes are comparatively recent (such as the ability to offer cabotage), and it is unclear in this situation of transition what the long-term outcome will be. There are, however, experiences from elsewhere and in other sectors that suggest reforms of the type carried through in the EEA may pose long terms challenges to the stability of the European airline market. In particular, there are concerns that airlines as a group will not be able to recover their full long-run costs.

Institutional changes

The most important institutional change that has occurred in Europe has been the phased liberalization of airline services both within the EU and within the countries that have linked into the EU air transport structure. This has opened up the intra-European market for airline services, and involved the EU in efforts to initiate more efficient air traffic control and to modify the ways in which airports have traditionally carried on their business.

A large part of these reforms have been initiated by specific air transport policies (the so-called 'Packages') but these have also been entwined with the applications of more generic EU wide policies in such areas as anti-trust. In the wider international arena, many European states have engaged in more liberal air service agreements, most notably in Open Skies arrangements with the US, that have broadened the basis of international competition.

Phased enlargements of Union membership that have been brought about under provisions of the Nice and Copenhagen Agreements, with the first being initiated in 2004. A major challenge for the future will be to ensure that these enlargements will be accompanied by appropriate air service provision.

These institutional developments have also been paralleled by other changes. There have been technology changes, including a move to using regional jets, the transformation of supporting telecommunications infrastructure, and investments in competing modes of transport – notably high-speed rail – have taken place. There have been shifts in the industrial structure of Europe, with a rapid expansion in service industries and in the growth in globalization.

Significant demographic changes are taking place as the populations of many European countries begin to age and the demands for leisure activities change. Business cycle effects, to which the air transport sector is particularly sensitive, have perhaps become more pronounced. Tied to this, the traditional international economic structure whereby at least one of the major world economies (Japan, Germany, and the US) have grown even if the others have been in recession no longer seems to hold. Japan and Germany seem to be afflicted with on-going structural economics problems. These features have added to the uncertainties that confront the air transport industry.

Issues of sustainability

To assess the prospects of the current structure of the European air transport system, and to assess whether it is economically sustainable, requires some notion of how the airline market functions. Indeed, this would also seem to be a necessary pre-requisite for any rational policy-making. In fact, there is still no firm consensus on what the natural market for air transport services is, even though a copious literature from academics, business analysts and policy makers has emerged during the last thirty years.

Numerous ideas have been floated, and sometimes supported for a brief period by empirical analysis, but most of these have proved to be transitory. To name a few, it has been argued that the market is naturally competitive, monopolistically competitive, perfectly contestable, imperfectly contestable, oligopolistic, unstable, and duopolistic. It may well be that there are various sub markets with differing features and requiring different policy approaches – for example, between the full service carriers and the 'no-frills carriers'. If this is so then it has implications for the larger development of air transport in Europe.

At present, however, given the inability of many carriers to recover their full capital costs there arise discussions about the need for consolidation of the industry. This implicitly assumes that the inability to recover costs is a matter of excess capacity given the current levels of demand. There are also secondary considerations concerning how such consolidation should be carried out. It may, for example, involve natural market attrition, with some carriers leaving the market or shrinking significantly in their operations. It may mean mergers, but it may also mean airlines coordinating their activities through stronger airline alliance structures.

The underlying thrust of the argument is that costs are too high and that excess capacity prevents scale economies of various types from being fully exploited. However, rationalization of this kind, when carried through in the previous severe downturn in the US, did not bring about a long-term equilibrium to that market. Indeed, even in the profitable period of the late 1990s their net profit margins averaged only 2.9%. What may prove more fundamental is the ability of airlines to generate sufficient revenues to recover their costs even if the capacity offered is not excessive.

Non-full cost recovery can, therefore, be viewed in a different way. It may be that there is no competitive structure that allows sufficient returns to be earned to cover all costs. In this case there are issues of subsidies and public service style obligations that come into the debate. But there are other possibilities that are often given less attention. These include giving airlines more market power, and with it the ability to raise more revenues from the consumer through the price structure and to control their capacity. Airlines already deploy a variety of strategies to both extract at least part of what customers are willing to pay (through yield management) and to stabilize their cash flows (through frequent flyer programs). But these seem to be insufficient for full cost recovery under the existing regulatory structure given the competitive pressures in the market.

The approach

This study begins by looking at the role air transport plays in economic development and political integration. For a variety of reasons, but possibly largely because it constitutes less than 1% of the EU GDP in national income accounts, air transport policy has tended in the past to have been relatively neglected. What is clear at the very basic level, however, is that whenever there is a disruption to the air transport system in Europe, be it due to industrial action by air traffic controllers or pilots, or the financial failure of a carrier, there is immediate and extensive public concern./Air transport is important to the citizens and business of Europe. /

To examine the sustainability of the current European air transport system, however, requires looking beyond simple aggregate national income data to consider exactly where aviation fits into the fabric of the EU economic and political structure, and the specific roles it can play.

To develop a sustainable air transport policy it is important to understand the key parameters of the industry and some of the current long-term trends. Airlines and cargo carriers are an important element in this picture and some appreciation of the nature of the market for EU airline services is necessary. Although there is no agreement on the exact underlying nature of the airline market, one can isolate some of the key forces that impinge on it, and some of the reactive actions taken by carriers in response. It is also important to try to separate from the short-term situation, the underlying structural changes that are occurring.

Formal, legal institutions are important. The development of EU policies, as they have affected various elements of the air transport sector, has clearly exerted powerful influences on current conditions in the airline market. The embodying of air transport within the CTP came late, but within a period of a decade changed the foundations upon which the sector rests. It is also helpful to appreciate that while there have been major formal institutional changes in the airline market, essentially changing it to one of free market entry and competition, these have generally been less pronounced in other segments of the overall air transport sector.

This provides a basis to outline and assess some of the stresses that are currently being felt in the air transport market and to assess whether the current institutional structure is robust enough to sustain an optimal air transport system within the EEA. The focus is primarily, but not exclusively, on the difficulties many carriers have in full cost recovery. It explores the possibility that there are inherent flaws in the premise that current levels of competition in the airline market are sustainable. But it also considers a number of other issues including the degree to which airlines are encumbered by non-economic controls and regulations.

There is a tendency, although not unique to air transport but certainly pronounced in it, to treat airlines differently. Certainly airlines as an industry do have their peculiarities but they also have features that are generic and found in other industries. Additionally, the European air transport market, while in some ways unique, has institutional and other features in common with other air transport markets. In both cases there are lessons to be learned. Consequently, there is some discussion of developments in other markets. But this discussion is tempered by the appreciation that lessons from the US domestic air passenger market, the one most often compared to that of Europe, inevitably requires caveats associated with them. There are many important differences between the US and the EEA.

The approach of this work is both to provide information about the long-term structural development of the European air transport market, and to examine some of the key challenges confronting the sector, and the various actors within it. Without an understanding of the underlying structure of the airline industry in Europe, and more generally, it is difficult to develop any commentary on the challenges that it faces over the long term. It is also important to understand how the current situation has developed in order to avoid re-encountering the pitfalls of the past and give a context to the industry's position at the beginning of the 21st Century.

The book

Each chapter in the book addresses a broad question. This provides the opportunity to look at the factors that currently influence the efficiency of European airlines and to see how the industry has moved to meet these challenges. It also allows for a discussion of the problems that were not always adequately addressed in the previous literature, as well as the newer ones that are emerging. This need is perhaps a reflection of the fluid world in which we now live and illustrates the need for industries such as air transport to keep abreast of larger developments.

The penultimate chapter continues this quizzical theme but rather than simply summarizing what goes before it focuses on more specific matters that would seem to have particular policy relevance. The concluding chapter then acts to summarize the material in the book.

The book is designed to be accessible. It is largely verbal in style with supporting tables and figures. There are no mathematics or theorems, although these can be important on occasions in aiding with understanding. Intuition is used where other work would offer firm proofs. Specialized jargon is unavoidable but there is a Glossary at the end that provides definitions of some of the key terms and concepts. It also provides the full titles of commonly used abbreviations and acronyms. There are also two Annexes that provide more details of the factual background to the European air transport market and set these within the wider international regulatory system.

Chapter 2

Do We Need Air Transport?

Introduction

Transport has traditionally been seen as a major facilitator of economic growth and vital for the creation of conditions of social and political cohesion. The Roman Empire relied upon a comprehensive road and seaport infrastructure to conduct commerce and to keep its domain intact. The Incas of South America carefully maintained mountain pathways for their runners to carry messages. Britain maintained its trade routes by constructing coaling stations around the globe. Initially the canals and railroads, and then the interstate highway system in the US were seen as integral parts of the country's modern economic and political development.[1]

More recently, the Treaty of Rome establishing the European Economic Community stipulated the need for a Common Transport Policy (CTP) and, subsequently, the European Union's (EU) establishment of Trans-European Networks (TENs) has demonstrated again the importance attached to transport in trade and economics within Europe. The United Nations has seen international transportation as a crucial factor in allowing the spread of the benefits of trade across the globe, and has established bodies such as the International Maritime Organization and the International Civil Aviation Organization to reduce institutional transport barriers.

Air transport networks are an integral and integrated part of any modern society. This is as true for Europe as for other parts of the world. Much of modern industry requires rapid, reliable and secure transportation of its people and of its goods. Just-in-time production is seen as an important key to efficient management. Individuals seek access to diverse locations at reasonable cost to enjoy their leisure time, to retain social contacts and to visit family within Europe and outside of it.

 EU

Air transport will inevitably be required to perform additional roles as the geographical boundaries of the EU are further expanded after 2004, and as globalization more generally takes place. Air transport also enables fulfillment of less tangible, but at least as important, social and political ties that can transcend narrower economic benefits.

The growing use of air transport over the past decades in itself points to its increasing importance. It reveals society's preferences in terms of its desire for

---

[1] There has recently been considerable economic debate initiated by the work of David Aschauer (1989) about the role of infrastructure more generally as a facilitator of economic development. For a survey and a discussion of this work, see Button (1998).

mobility and interaction. The long-term forecasts of future air travel demand confirm the continuation of this pattern. In addition to its global and national implications, access to high quality air transport services has important regional and local effects, and is key to the success of many industries. Modern society often places as much emphasis on spatial equity as on overall national income levels.

The demand for air transport

Economists consider that in the short term the demand for air transport services is a derived demand. People and companies use air transport to meet some other need rather as an end product in itself. The use of charter flights in this sense can be seen as way for people to enjoy the end product of a vacation. As we see below, however, the availability of air transport can also act to enhance economic growth as well as being a follower of it.

In terms of its economic characteristics, air transport is a mature industry in the sense that it offers reliable, safe, and relatively low cost services. It is a large industry. While there is some fluctuation over time about the long term trend, it is also a growing industry. Globally, it carries some 1,600 million passengers a year, is responsible for 3.9 million jobs, has a turnover of $260 billion, engages 18,000 aircraft, and serves a 15 million kilometer network involving 10,000 airports. Air cargo traffic is over 130 billion revenue tonne-kilometers.

Air transport is the dominant form of transport over long distances but still serves the needs of many middle distance travelers (Figure 2.1). The average route length in Europe is 720 kilometers, but is 1220 kilometers in the US, where the major centers of economic activity are more dispersed.

Perhaps more important than the crude numbers are the conclusions of the US's National Commission to Ensure a Strong Competitive Airline Industry (1993) and the European Union's (EU) Comité des Sages for Air Transport (1995). Air transport is not only a major industry but is of considerable significance as an input into rapidly growing national, international and global economies. It is an essential ingredient for the success of tourism in many regions. It is also a vital input in the development of numerous, non-leisure-based industries where interpersonal communications are important.

It is not only passenger air transport that is vital to many of these industries. Many firms also rely on air freight to provide quality service to customers. Air transport is an important link in many supply chains, and is often essential for the effecitive operation of just-in-time production management.

Market liberalization

To facilitate improved efficiency in air transport supply there has been a trend towards greater market liberalization over the past twenty years. Most air transport markets were heavily regulated from their inception, with controls over fares, service levels, revenue allocations, and suppliers. The US broke with this mold in

1977 when it liberalized domestic cargo markets, followed a year later by the deregulation of its domestic passenger market.

In recent years, there has also been a movement away from regulation in the international market. Blocks of nations, such as those in the EU, have developed multilateral free markets among themselves. Further, many bilateral agreements between countries governing the terms on which air services are traded, have been liberalized. The US's Open Skies policy of the post 1979 era, but more especially from the mid-1990s has, for example, led to much greater openness in many international markets, but controls still remain over significant parts of the global network.

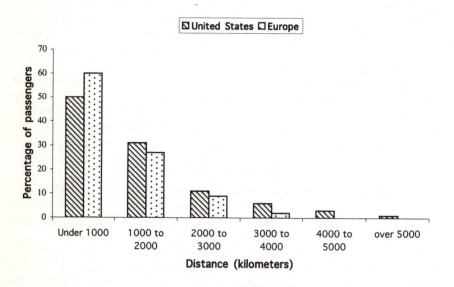

Figure 2.1 Passenger numbers by distance of service

Ultimately, in democratic societies, the importance of any activity comes down to whether, if the full costs of its provision are being recovered, people wish to buy it. Industry analysis reported in later chapters indicate that imperfections in the value chain lead to higher than necessary costs for airlines. But despite this passenger kilometers in the EU grew on average by 7.4% per year from 1980 to the end of the century while the traffic handled at the major airports of the Union increased five-fold.

Forecasts of future demand

Traffic forecasts by major interested companies such as Boeing, supply groupings, such as IATA, and international agencies such as ICAO, all indicate that society will want to consume even more air transportation services in the future. Such forecast should, however be set within their particular context. First, they are all context specific which influences their focus – Boeing, for example, is primarily interested in the future demands for airframes Second, short-term air traffic forecasts are notoriously inaccurate, but longer-term predictions, with a slight bias to the conservative, have been more reliable with time. They do seem to offer a broad indicator of the attitude that society is taking regarding the importance of air travel and air cargo.

Medium term forecasts of international passenger traffic from the International Air Transport Association (IATA) conducted in 2002 and deploying a Delphi method of analysis to elicit the views of some 280 airline officials. This entails initially posing questions regarding future prospects to the subjects and then allowing them to refine their responses after learning of their competitors' views, The study found significant differences in the ways various segments of market are likely move in the future; intra-European and European Asian markets having the greatest short term potential.

Both the IATA and the ICAO, however, think the overall market for air travel will exhibit short-term positive growth trends (e.g., Figure 2.2). These trends can be seen as reflecting the importance attached to mobility despite the higher security costs and difficulties of travel that the post September 2001 situation demands. Whilst the subsequent impact of SARS on several markets, and the effects of the Iraq conflict have held back recovery, the overall pattern seems to be generally accurate. The long-term tend is also generally upward.

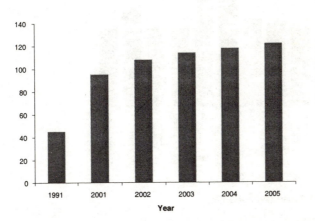

Figure 2.2 ICAO short-term forecast of passenger traffic growth

Longer-term forecasting of air transport markets, even of broad trends, is inevitably difficult. In effect the approach involves some fairly complex extrapolations to produce its predictions. This means making assumptions about trends in income, fares, costs and other determining variables. Added to this, there is a need to assume that the relationships between these factors and the demand for air services remains the same through time. It also normally involves looking at hard data and trends, and making used of econometric modeling techniques, rather than considering expert opinions in the manner of the IATA analysis.

With these caveats, forecasts from Boeing Commercial Airplanes (2002a; b) predicts annual growth in revenue passenger kilometers will be of the order of 4.9% over the next 20 years. This is about 2% points above the projected growth in global GDP over that period.

Spatial variations in markets

There are important regional differences in the forecast growth rates of revenue (Figure 2.3) that reflect a combination of factors such as the maturity of the various markets and their geographical features. If accurate, these forecasts would indicate that Europe would be in the middle range of growth markets, although that growth will be approximately double the rate of anticipated economic growth. Projections by the EU indicate that by 2010, 8% of European passengers will travel by air.

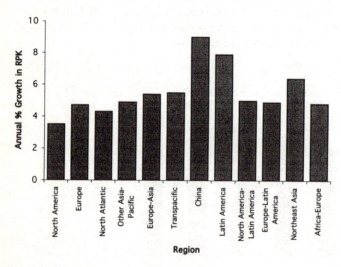

Source: Boeing, *Current Market Outlook* 2002.

Figure 2.3 Long-term forecasts of revenue passenger kilometer growth to 2021

Air cargo traffic is highly dependent on the growth in global GDP and in the structural changes that are taking place in many markets. The forecasts from Boeing are that the global annual growth rates to 2021 will be somewhere between 5.1% and 7.8% with 6.4% being the best estimate. If this materializes it will result in 464 billion revenue ton kilometers being done by 2021. The intra-European market will remain small, but there will be important increases in the Europe-Asia market and the Europe-North America market, both of which will grow, according to the Boeing Company, faster than the global average.

The economic benefits of locating near a major airport

One of the most immediate ways of gaining an insight into the importance of air transport to society is by considering the implications for a local economy of having good air transport services. There are some convenient studies available that throw light on this. The impacts of air transport are regularly demonstrated in the impact analysis done to justify the building or expansion of an airport, or when secondary construction is undertaken, for example to improve road access to an airport.

Quantitatively the economic impacts of having good air transport access depend primarily both on the time frame examined and the geographical space under review. In broad terms, airports have four potential economic impacts of varying duration and spatial coverage (Table 2.1).

Table 2.1 Economic impacts of airport development

- *Primary effects.* These are the benefits to a region in the construction of an airport – the design of the facility, the building of the runways, the construction of the terminals and hangars, the installation of air traffic navigation systems and so on.
- *Secondary effects.* These are local economic benefits of running and operating the airport – employment in maintaining the facility, in handling the aircraft and passengers, in transporting people and cargo to and from the terminal and so on. These secondary effects can be extremely important to some local economies in terms of employment, income and, for local government, taxation revenue.
- *Tertiary effects.* These stem from the stimulus to a local economy resulting from firms and individuals having air transport services at their disposal. These differ for those living in hub cities, compared to those on a spoke or having no major carrier. Hubs offer more direct flights favored by business travelers. But the hub also benefits those on the spokes because without a hub-and-spoke structure many would find it difficult to travel long distances at all. Hubs allow interconnectivity. In the US over half of the 15,000 city pairs served by a major carrier have less than one passenger per day.
- *Perpetuity effects.* These reflect the fact that economic growth, once started in a region, becomes self-sustaining and may accelerate. An airport can change the entire economic structure of a region – it can shift its production function. This type of dynamic economic impact of an airport is the most abstract and the most difficult to quantify. It has been little researched.

Whilst it is interesting to isolate these beneficial effects, there remains the practical question of how large are they? It is also often challenging to isolate the individual components.

Each airport is unique and as such the scale and nature of its economic implications for the local community is highly case specific. There are several ways in which linkages can be teased-out, each has its merits and limitations. It can be done through questioning those involved in running the airport or are involved in local industry, but this poses problems of ensuring objectivity in the responses obtained. It can be done using local Keynesian multipliers or input-output analysis, but there are then issues of the geographical coverage and the time frame to consider. Finally, there are econometric methods that make use of statistical techniques but while offering the ability to isolate air transport effects on local development, do pose problems of appropriate model specification and estimation procedures.

But irrespective of the method used, the body of evidence on the potential scale of secondary and tertiary effects of air transport provision on economic growth is compelling (Table 2.2).

Pulling the information that we have into standard economic impacts is not easy because of the differing circumstances surrounding each investment, but some general impressions can be gleaned from the studies cited above and other work. This type of information is summarized in Table 2.3. The table provides a set of guidelines as the impact of having a relatively small airport (a million passengers a year) in a community in terms of job and local income creation. It offers a range of estimates that will be influenced by such things as the types of services offered and the nature of the surrounding economy.

Tourism

It is perhaps a trait of humans that they actually like to visit new places. Tourism has been around at least as long as people visited the Seven Wonders of the World. The Grand Tour was part of a gentleman's 'finishing' in the 18th century. Mass tourism began with Thomas Cook and the use of the railway system for leisure travel. The rising incomes and increased amounts of leisure time enjoyed by many people in industrialized countries in the latter part of the 20th century, combined with enhanced transportation systems, led to an expansion in international travel for recreational purposes.

Each year there are seasonal mass short-term migrations to areas where the climate is more favorable, the scenery more attractive, where there are different cultures to sample, or where there are antiquities to see. Added to this, families are becoming more spatially extended and as a result family visitations often involve longer trips. As a consequence, tourism (and the visiting of kith and kin) is now a major industry that is a significant source of income and employment in many developed and developing economies.

Table 2.2 Examples of airport impact studies

Survey techniques
- The Atlanta Chamber of Commerce found from a survey of 264 foreign-based firms, that availability of direct international services was the third most important factor in location decisions. Subsequent study showed that the number of foreign firms locating in the region from a particular country grew significantly after the introduction of a non-stop service.
- Ernst and Young, looking at location decisions of 57 companies in Europe making decisions regarding the location of a manufacturing plant found that the air transport network was the third most important factor in the decision process. Air services were much more important for service sector companies.
- The Amsterdam Chamber of Commerce found that the availability of an airport was one of five key factors considered in company relocation decisions.
- A survey of firms around Zurich found that 34% considered the airport as 'very important' and 38% as important as a location factor.
- Loudoun Chamber of Commerce (Virginia) found that airport/freeway access was important to 68% of firms.
- A study of small business firms in the Washington area that were engaged in export activities found the availability of easy access to international air transport one of the six most important factors in their success.

Multipliers
- An academic study by Rietveld estimated that Schiphol Airport (Netherlands) generates about 85,000 jobs for the country.
- A study of Vienna International Airport by Industriewissenschaftliches Institute in 1998 indicated that on a turnover of ATS25 billion in 1996, there was an impact of ATS11.2 billion on the local economy.
- The Institute of Social and Economic Research found that the total annual economic importance of Anchorage International Airport on local payrolls was $130 million above the $316 million for on-site activities.

Econometric
- Analysis by George Mason University taking variations in high-technology employment across all US Metropolitan Standard Areas, found that a hub airport in a region increases that region's new economy employment by over 12,000.
- Brueckner looking at possible expansion of Chicago O'Hare airport found that an increase of traffic of 50% will increase service related employment in the region by 185,000 jobs.
- An econometric study by Science Applications International of the implications of the Open Skies agreement between Germany and the US on the regional economy around Hamburg Airport found annual gains for the regional economy of $783,318 in 1994 prices.
- Button and Taylor examined the impact on US cities of having European services and found that employment was systematically positively related to both the scale of services offered by airports and the range of destinations served.

Table 2.3 Jobs and income from having an airport (per million passengers)

Estimate	Jobs		Economic Impacts ($ millions)	
	Direct	Total	Direct	Total
High	2000	8000	225	1600
Medium	1500	6000	75	650
Low	750	2500	35	130

The tourist market

The scale of the modern tourist industry is large. It is a global industry. It takes a diversity of forms. The World Travel and Tourism Council (WTTC), the tourist industry's main lobbying organization, has argued that the travel and tourist industry is the world's largest industry and it is growing rapidly – globally, expenditure on tourism has been rising at an average of about 5% per annum since 1970. (The number of tourists fell in 2001, following the attacks in the US, but prior to that it had risen every year since 1982.)

In all regions of the world, excluding Europe, air transport is the major mode for transporting tourists – in Europe shorter-distance vacation travel is more common and the automobile accounts for much of this.

Whilst the details of many calculations surrounding the tourism industry are open to debate, it is certainly true that globally tourism is a significant employer and generates large amounts of income. Of more importance are its roles in particular contexts and locations. Many small islands in Europe depend on tourism as their main source of income and their dominant source of foreign exchange. Even in larger economies, tourism can play an important role in generating employment and income (e.g., it accounts for 12% of Spain's GDP) as well as income for transfers elsewhere.

On the supply side, the originating regions are still very concentrated with 58% of international tourists coming from Europe and 18% from the Americas; indeed 80% of all international travelers originate from only 20 countries. It seems likely that the traditional destinations will continue to take a large share of the market. For example, while the gap is closing, Europe received some 717 million tourists and East Asia and the Pacific some 397 million (Table 2.4), Europe is still likely to retain its position as the main tourist destination in the foreseeable future.

Despite the increased demand for longer distance tourism, intra-regional travel still dominates. The most popular destinations for international tourists in 2000 were France (75.5 million), US (50.9 million), and Spain (48.2million). Changes are taking place however, and by 2020 it is forecast that China, with 186.6 million visitors, will be by far the world's largest tourist destination.

Table 2.4 Forecasts of tourist growth ($millions) by destination

Destination	1995	2020	Average Annual Change
Europe	338.4	717	3.0
Americas	108.9	282	3.9
East Asia/Pacific	81.4	397	6.5
Africa	20.2	77	5.5
Middle East	12.4	69	7.1
South Asia	4.2	19	6.2
Intra regional	464.1	1,183	3.8
Intercontinental	101.3	378	5.4

Source: World Tourism Organization.

The economics of tourism

In financial terms, estimates from the WTTC indicate that tourism in total (including domestic tourism and travel) contributed $69 billion to the UK's economy and $68 billion to the French economy in 2000. WTTC estimates are based on personal consumption, business travel, visitor exports and government expenditures, such as subsidies paid towards cultural attractions, parks, and museums. In terms of gross foreign exchange earnings, Spain gained $33 billion and France $30 billion, while the US was the top spender on foreign tourism ($59 billion). Isolating international tourism, the global amount spent was $476 billion, or about $680 per international tourist. Some 59 countries belonging to the World Tourist Organization enjoyed receipts amounting to $1 billion or more in 2000. In total, tourism accounts for about 32% of world exports of commercial services

The scale and geographical diversity of modern tourism and travel would not be possible without advances in mass transport. In some cases, such as tourism involving islands, the advent of cheap air travel and cruise-liners was a necessary precursor. But even mainland tourist destinations, such as the theme parks in the US and the beaches of the Mediterranean, only became mass tourist destinations because of air transport and the growth in motorcar ownership. The growth of low cost airlines ('no-frills carriers' in European markets) has also led to increased numbers of short breaks or weekend holidays in some markets.

Airlines serve an important role in moving people to tourist destinations (Papatheodorou, 2002). In 1998 air transport represented 43% of international tourist movements with road transport accounting for 42% and over time the importance of air has been growing.

There are important regional differences. For example, road and rail account for nearly 60% of international trips in Europe since it consists of many small countries. The data is also somewhat distorted because the definition of a visit is based on a minimum of a one-night stay and, hence, a car trip from Denmark to Spain with one-night stops in France each way en route counts as three tourist trips (two in France and one in Spain), flying counts as one (Spain). But for particular areas, air traffic's share can be much higher – e.g., 80% for some parts of the Mediterranean region.

The European market for tourism is also different from that of other large markets such as the US. In Europe, tourism involves much sight seeing and resort activities (Figure 2.4) that inject significant money into tourist locations. In the US, tourism more commonly involves visits to friends and relations suggesting local expenditures are lower.

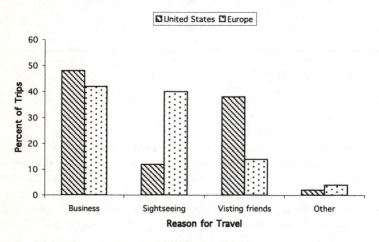

Sources: Institute of Air Transport, Gallup/ATA Survey.

Figure 2.4 Motivations for travel in the US and Europe

A large part of the European tourist market has traditionally been served by non-scheduled operations. Charter operations do about 44% of the intra-European revenue passenger kilometers and carry about 27% of passengers – the difference being due to the non-scheduled services providing carriage on the longer north-south routes (Table 2.5). Specialist charter carriers carry many of the passengers but scheduled carriers such as Finnair, Lufthansa and Turkish Airlines are also significant players.

Air cargo

While much public attention focuses on airline passenger services, modern air transportation began as a postal service, and it continues to fulfill the role of a cargo carrier. Air cargo, which encompasses the carriage of mail and packages, is a large (Figure 2.5), modern, and very rapidly growing industry that provides vital services to a variety of commercial and industrial sectors. Globally, approximately 30 million tons of cargo is flow annually, accounting for some 130 billion revenue ton kilometers. The sector grew by over 6.5% per annum worldwide throughout the 1990s. While there have been shifts in markets over that period, Europe-Asia and Europe-North American markets remain as two of the largest global markets, and are forecast to retain that position.

Table 2.5 Major international charter routes in the European Economic Area

Between	Percent Charter	Total Passengers
UK/Greece	81	6.5 million
UK/Spain	80	24 million
Austria/Greece	80	1 million
Finland/Spain	76	0.4 million
Denmark/Greece	93	0.65 million
Germany/Greece	75	5.2 million
Netherlands/Spain	53	3.3 million
Ireland/Spain	86	1.2 million
Norway/Spain	92	0.86 million
Sweden/Spain	91	1.7 million

In part because of formerly rigid bilateral air service agreements, the large national flag carriers (Lufthansa, British Airways, and Air France) have traditionally dominated the European market for cargo carriage. With economic liberalization, and shifting synergies between passenger and freight carriage, the integrated carriers have been taking an increasing share of the market over the past decade – they are now performing about 44.0% of the intra-Europe revenue ton kilometers.

The European cargo suppliers, however, with the exception of Lufthansa Cargo, are still relatively small compared with their large US counterparts such as FedEx – 14,632 million revenue tonne-kilometers (RTKs) and $16,351 million in revenue in 2002 – and UPS – 7,295 million RTK and $2,846 million in revenue. In aggregate, European airlines in 2002 accounted for about 35,000 million RTK out of a global total of 143,000 million and for about $9,000 million of global air cargo revenues of $51,000 million.

For most European airlines, and especially the former flag carriers, that serve both passenger and cargo markets, cargo provides a useful, if not dominant, source of revenue (Table 2.6). Compared to the RTKs done, it provides a lower rate of return than does passenger transport, but if passenger activities are the primary interest cargo carriage may be seen as an important contributor of marginal revenue.

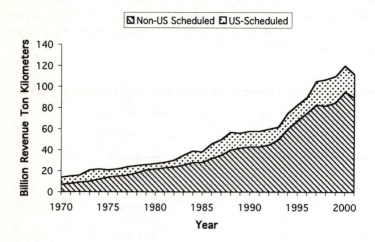

Source: Boeing, *World Air Cargo Forecast*, 2002-2003.

Figure 2.5 The global carriage of cargo by scheduled air freight

Table 2.6 Air cargo by European carriers in 2002

Airline	Country	Cargo (RTK)	Revenue ($ million)	Cargo share of: RTK	revenue
Lufthansa Group (3)	Germany	7,158	2,191	45%	14%
Air France (7)	France	4,862	1,480	34%	12%
British Airways (10)	UK	4,210	752	30%	6%
KLM (11)	Netherlands	4,197	1,018	41%	16%
Cargolux (12)	Luxembourg	4,157	794	100%	98%
Matinair (19)	Netherlands_	2.465	381	78%	51%
Alitalia (30)	Italy	1,378	397	32%	9%
Swiss (36)	Switzerland	1,028	273	51%	10%
SAS Cargo (37)	Denmark	928	236	29%	3%
Virgin Atlantic (39)	UK	894	154	26%	7%

Note: Figures in parentheses are world ranks by RTK.

Air cargo transport is not as important in terms of the tonnage involved as in terms of the nature and value of the goods moved and the lubrication function that it serves in the modern chain of production. Indeed machinery parts are the largest single category of goods moved by air cargo. It is usually the fastest way of moving goods and packages over long distances, and provides the backbone of such activities as express mail services. It is also a very secure mode of transport because of the isolation of the plane during a flight – a not unimportant consideration in the early part of the 21st century (Coughlin et al, 2002).

Because of the quality of the service offered, to meet particular consignor needs, air cargo serves a number of key functions. This has resulted in a degree of market segmentation. The heavier cargo movers provide airlift for high value products where speed of delivery (newspapers and medicines) and security and safety are important considerations. They also provide a major form of long-distance transportation for perishable goods, most especially seafood, fruit and flowers. This has enabled, for example, the globalization of the Netherlands' flower industry and also permitted many lower income countries (notably in South America and Africa) to market their products in Europe and, via integration activities, in other markets.

The advent of sophisticated supply chain and inventory management has put additional demands on air cargo. The widespread adoption of just-in-time production techniques has reduced the holding by companies of inventories and this has resulted in the need to provide components in a predetermined and assured manner. In the case of high value, low weight inputs air transport has a major advantage over surface based modes. Airfreight is also a key player in transporting spare parts when there are technical breakdowns and in the rapid movement of documentation.

In recent years the large express carriers (FedEx, UPS, etc) have, through their direct supply of services, been a competitive stimulus for the more traditional postal systems. Their integrated transport systems, combined with modern communications networks and distribution algorithms, have facilitated the growth of E-Commerce and provided a core component to the creation of service-based economies. Their hubs in the US (e.g., Memphis for FedEx), and to a lesser extent in Europe (e.g., Frankfurt, Paris, Schiphol and London) have become major commercial interchange points with local industry offering value added services.

The intra-European express sector grew rapidly in the late 1990s (by nearly 30% a year from 1989 to 1999 on a ton-kilometer basis) and, although hit by the recent effects of the slow economic growth of many European economies, has managed to run counter to many other air transport markets by expanding by 1.4% in 2000 and 2.2% in 2001.

The general view is that air cargo will continue to grow in the future with much of the additional traffic coming from the freight side. Airmail is likely to grow consistently at about 3.0% per annum according to Boeing's forecasts. The development of electronic communications systems is seen as offering increased competition to paper documents, but there are also complementary effects at work. As telecommunications is taken up in more applications, this stimulates transactions more broadly, including those involving hard copy movements.

In contrast to the staid growth in airmail, freight is predicted to grow at between 5.2% and 7.9% over the next twenty years as industrial production changes and as increased spatial specialization materializes. The intra-European market is forecast to grow at best as the lower end of this range, although express networks will grow faster. The geographical distribution of this growth within Europe will be strongly influenced by the economic performance of individual economies

Chapter 3

How is the European Air Transport Market Developing?

Regulating the air transport market

The aim of this chapter is to look at the broader and longer-term changes that have been taking place in European air transport markets. It offers what is often called a 'meso' level of analysis in that the focus is neither on individual carriers nor routes but is at a higher level of aggregation that is centered more on industry and larger markets. Chapter 4 then moves on to provide a more detailed look at the on-going segmentation of the European market, and in so doing also focuses a little more on specific carriers.

We begin by considering regulations, after all institutions are important in determining how markets function. The initial rationale behind domestic economic regulation of air transport had much to do with enhancing mail services. This philosophy was later extended to passenger services with the goal of ensuring adequate transport provision to regions that were in need of enhanced access. International traffic was initially regulated on an ad hoc basis and often designed to meet strategic objectives. The development of Empire routes by colonial powers was a clear manifestation of this.

Of considerable political and institutional importance was the standard approach to the regulation of international services that followed the Chicago Convention of 1944. This gathering to define the post-Second World War structure for international air transportation codified aviation rights and permitted the creation of a freer international market. In particular the matter of national sovereignty was clarified. This gave the right to each country to regulate air services into and out of its territories. It established the concept of 'Freedom of the Skies' that offers a structure for further institutional developments.

While the overall picture was clarified at Chicago, the detail was not resolved. As a result, in practice progress after 1944 has been slow and the largely bilateral structure that emerged produced a restrictive regime albeit less so than the ad hoc system that went before.

Countries regulated their air transport market largely to protect their national airlines. Beyond strategic considerations, they were seen as important vehicles for furthering a nation's prestige and commerce. In economic jargon their existence had external benefits beyond those of simple commercial viability. This type of rationale has evaporated as air transport has become much more of an everyday activity and it has become appreciated that it can often best serve the external interests of a country by being provided commercially. In many cases European states had only a single international carrier designated under a bilateral air service agreement. This gave the carrier a certain amount of market power. In the US this was handled through regulation of private sector suppliers but in most European countries flag carriers were publicly owned. The situation has now significantly changed.

The recent moves to liberalizing airline markets may be seen as a reaction against these types of regime. But regulations of a more generic type still persist. These come under the guise of mergers policies and competition policy (anti-trust policy). They follow the broadly accepted argument that competition is inherently desirable and that market power or cartelization is potentially inefficient – leading to high prices, limitations on supply, and a lack of innovation. Regarding air transport, such policies have been applied to mergers, alliances, and predatory behavior.

The difficulty with these arguments is more in the application of their ideas than the underlying theory itself. For example determining market power requires the definition of a market that has clear boundaries. This is particularly difficult in network industries such as air transport. While only one supplier may offer a service from A to B, another may offer an indirect service through C – does that constitute competition?

Predatory behavior is usually seen as a situation when an incumbent drops price and/or increases output to combat a new competitor, and in so-doing forcing the newcomer from the market. But when should this be treated as predatory rather than just being a natural element of healthy competition? The difficulty of these issues, coupled with the natural carefulness of judiciaries means that decisions on these types of matter can take many months as witnessed by recent judgements on a number of airline alliances.

Trends in air transport policy

Globalization and internationalization are two of the most pronounced, not to say controversial, industrial trends of the late twentieth and early twenty-first centuries. Their emergence has resulted in more spatial specialization of production, a growth in trade, more mobility of capital, and movement of labor. The driving forces for these developments are not altogether certain but regulatory changes, new production technology, and lower transport costs are unquestionably components of the interactive mix that underlies them.

Part of the linked phenomena of globalization and internationalization is reflected in the significant trade growth that took place in the 1990s, with real export growth in the Organisation for Economic Cooperation and Development (OECD) area at over 7% per annum. It is premature to judge whether these trends are passing fads or genuine long-term adjustments to the way production and trade is conducted, or indeed whether socio-political conditions will act as an ultimate break.

Market liberalization

These developments have been taking place whilst the institutional structure of European air transport services has seen significant change. Institutionally, the intra-European market has moved to a situation akin to that found within the US. Many European countries unilaterally liberalized their domestic airline markets in the early 1980s. But it was the EU that, since 1988 through a succession of 'Packages', moved European airline markets to a position that leaves airlines largely free from economic regulation.[1]

The creation of a Single European Market in the 1990's meant that international air transport within Europe was essentially deregulated, with full cabotage within member states being permitted from 1997. This does not mean that there are no regulations governing airline activities, but rather what is conventionally called economic regulation (controls over prices and capacity) have been removed.

This is not a uniquely European trend. US domestic markets were liberalized from the late 1970s. A majority of markets in South America have been liberalized with various types of privatization programs. Australia and New Zealand markets have been deregulated. Additionally, the World Trade Organization (WTO) brought into play (albeit an extremely small role) a new and geographically wider policy making institution to supplement the roles already played by bodies such as the International Civil Aviation Organization (ICAO) and the IATA. Aviation issues are on the agenda of new regional groupings such as the Asian-Pacific Economic Council (APEC).

These institutional reforms, at a time of rising incomes and increased leisure time for many, have contributed to the steady growth of demand occurring in aviation markets. Additionally, technology advances have allowed aircraft efficiency to rise – albeit it at a very much slower rate than in the 1960s and 1970s – and air traffic control and navigation systems to handle more traffic. Short term market retrenchment (often a slacking of growth rather than a decline in traffic) regularly accompanies downturns in the business cycle. Serious adverse effects were seen as a result of the 'shocks' such as Gulf War and the attacks on the US in September 2001. These reflect, as in most markets, short term mismatches of supply and demand.

Since 1960, air passenger traffic has grown globally at an average yearly rate of 9% (Figure 3.1) and freight and mail traffic by some 11% and 7% respectively. Some 1.56 billion passengers per annum were being carried by the world's airlines at the close of the 20th Century. Association of European Airlines (AEA) members saw a steady growth in their traffic in Europe during the 1990s that produced a doubling of passenger kilometers done. By 2000 the market was 32 times the size of 1960. In addition to passenger transport, aviation is an important form of freight transport, with some estimates suggesting it carries up to 40% of world trade by value and is forecast to rise to 80% by 2014.

---

[1] Annex I provides more details of the development of EU air transport policy.

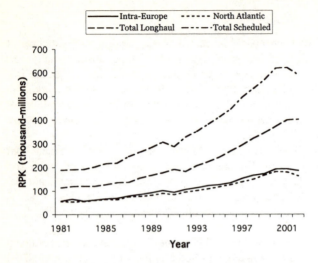

Source: Association of European Airlines, *Yearbook, 2002.*

Figure 3.1 Long-term trends in European based passenger markets

Fares and cargo rates

One factor exerting a positive influence on these trends has been the steadily falling average fares for passengers and lower freight costs for air cargo in Europe over the past decade as the phased liberalization has gradually impacted (Figure 3.2).

Much of the downward trend in air transport costs is generally attributable to technology advances prior to the 1980s (e.g. the introduction of jet engines and then of wide bodied planes) with regulatory reform taking a role thereafter. But there have also been accompanying managerial changes that have enhanced productivity. The remarkable increase in European air transport in recent years, for example, has been achieved with a much smaller relative increase (6.0%) in the labor employed by airlines in the decade since 1992 (Table 3.1). The larger employment in the European air transport industry more broadly (which also includes airport, ground handling and catering employment) rose from 435.4 thousand in 1988 to 489.7 thousand in 1996.

Revenues and profits

The profitability of airlines in Europe has varied across carriers and across the business cycle. After running a deficit for several years in the early 1990s, most airlines managed to make profits in 1995. Net profits for the 12 main EU airlines were in the region of US$800 million against a net overall loss on the same scale in 1994. However, financial performances varied, with only British Airways, Finnair and KLM achieving universally favorable results over the entire period from to 1994.

Source: Association of European Airlines.

Figure 3.2 Real passenger and cargo yields for European scheduled services

Table 3.1 Employment by airlines in Europe

Country	1992	2001
UK	70,838	98,649
France	60,583	69,235
Germany	58,168	51,979
Spain	30,594	34,492
Scandinavia	28,733	31,173
Netherlands	28,364	31,459
Italy	23,790	25,219
Switzerland,	22,813	14,535
Portugal	12,019	9,809
Belgium	11.043	1,542
Greece	10,861	5,625
Finland	8.053	9,240
Austria	5,566	7,536
Ireland	5,453	6,323

Among the medium-sized and regional airlines that were particularly active since the introduction of the Third Package of EU reforms in the late 1990s, Regional Airlines and (the now defunct) Air Littoral had net profits of Ffr9 million and Ffr8.5 million respectively in 1995, with EBA making net profit of Bfr200 million and Tyrolean Airways, $3 million.

The later years of the 1990s saw profitability rise across the main European carriers, with aggregate profits on international routes exceeding $2.25 billion for 1997 and 1998. This mirrored wider international experiences. Higher fuel prices combined with enhanced competition in the late 1990s squeezed profits resulting in aggregate losses in 1999. In particular yield, which had fluctuated at around $0.04 to $0.05 revenue passenger-kilometers from the late 1980s to the mid-1990s dropped significantly and by 2000 was below $0.03. The result was that operating margins began to fall sharply (Figure 3.3).

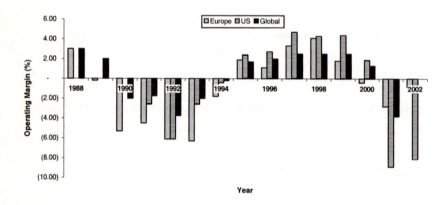

Note: Omitted years indicate missing data not a zero margin.
Sources: The US data refers to airlines that are members of the ATA and the European data to airlines that are members of the AEA (financial years). The global data is from Boeing Commercial Airplanes.

Figure 3.3 Operating margins after interest as a percent of operating revenue

The onset of economic recession in the US in 2000 and the events of September 2001 pushed operating margins dramatically into the red. Financial recovery has subsequently been generally weak as the performance of the world's major economies (notably the US, Japan, and Germany) have been sluggish. The shock of SARS on the transport market and the invasion of Iraq have added to the depression in the market.

The situation is not unique to any single geographical market. Figure 3.3 also details the net operating margins of major US and EU passenger airlines and globally. Different accounting conventions, financial years and airline composition preclude a direct comparison but the broad picture for the 1990s is clearly very similar. However, forecasts (see again Chapter 1) indicate that, as a sector, aviation will continue expanding in the foreseeable future, albeit at different rates, and with variations across geographical sub-markets.

The market for airline services

In addition to market forces, a variety of other factors influence the shape of European air transport. The underlying nature of the airline market is relevant but so too are the institutional arrangements that shape the parameters of this market. Institutions in the European context do not merely embrace rules and regulations that impact on domestic aviation, but also reflect the bilateral arrangements that exist between each pair of countries and the multilateral arrangements, such as those within the EU, that have been established.

Non-market forces

International, EU, and national institutions are important in shaping air transport, and take a wide variety of forms. There have in the past been a number of international agreements affecting the way the EU air transport market functions. Some were drawn up between EU member states and have been significantly changed in recent years, but others transcend EU borders. These involve legal and regulatory structures that have grown up over many years, often designed to meet economic and political challenges of days now long gone.

There are important extra-EU institutional arrangements that have implications for aviation in the Union, because a lot of air transport within the EU actually originates or is destined for countries external to it. Also important are non-air transport institutions, such as general competition laws, that influence the structure of the air transport supply.

These institutional structures, and the de facto ways in which they function, have not emerged in a vacuum. They are products of a variety of factors, not least prevailing political, social and economic theories and the experiences of previous policies. They can also be affected by developments outside of the EU, for instance experiences of regulatory reforms elsewhere, that exert an indirect 'demonstration effect' on European thinking and actions.

Some of the diversity of forces affecting European air transport can be seen in Figure 3.4. They are discussed in more detail elsewhere (Button et al, 1998) and are only given cursory treatment here. What was perhaps of often forgotten underlying importance in the 1990s was that while many of these forces were driven by pure pragmatism, for example the need to reassess national air transport regulations as changes were occurring in air transport markets and in regulatory structures more widely, there was also a genuine intellectual concern to improve institutional arrangements.

The ideas of economists in particular were shifting regarding the role of institutions, and the ways in which industries are best regulated. A gradual skepticism had emerged in the 1970s, especially in the US, about whether regulations always served the public interest, or whether they more often served the interests of the regulated or the regulators. Were the failures of markets indeed that bad compared with the potential failures of inappropriate government interventions in the market through ill-conceived regulation or its manipulation to serve specific interest groups?

But it was not all ideas. Prior to changes in US regulatory policy a body of academic work comparing regulated US inter-state air transport markets with more liberal intra-state regimes, notably California and Texas, had emerged (see Annex II). The latter were clearly offering lower fares. The US experience with airline deregulation itself from 1978 offered strong demonstration effects of the implications of more liberal markets, as had other earlier reforms within the EU such as the deregulation of the UK road haulage industry under the 1968 Transport Act.

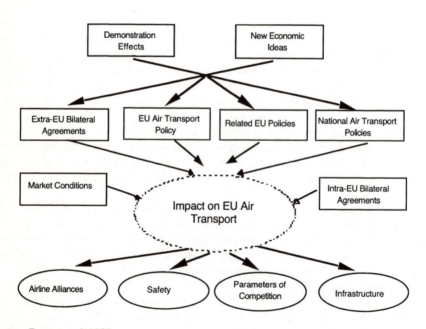

Source: Button et al, 1998.

Figure 3.4 Simplified flow chart of factors influencing EU air transport

Scale effects

European airlines differ in terms of their sizes and the extent to which they engage in services outside of Europe. European airlines are also relatively small compared to their US counterparts. The US has the world's three largest passenger airlines (American, United, and Delta) in terms of persons carried, with Northwest and Continental Airlines also in the top 10 (Table 3.2). All of them are active on transatlantic routes. Only Air France (5th), British Airways (6th), and Lufthansa (7th) are in this grouping. FedEx Express is also the world's largest cargo airline.

The picture is slightly different if airlines are ranked by financial criteria, but the size of the US carriers was still pervasive in 2002. The integration of KLM (ranked 13th by revenue) with Air France will, however, produce some change in the rankings in future.

Table 3.2 The ten largest airlines in 2002 ranked by revenue

Airline	Country	Revenue ($ million)	Passengers (million)
AMR Corp /American Airlines	USA	17,299	15,772
Japan Airlines System Corp	Japan	17,244	10,741
Lufthansa Group	Germany	16,123	9,240
United Airlines	USA	14,286	11,872
Delta Air Lines	USA	13,305	12,321
Air France Group	France	12,697	10,535
British Airways	UK	11,940	10,165
ANA Group	Japan	10,063	6,246
Northwest Airlines	USA	9,489	8,025
Continental Airlines	USA	8,402	7,862

Source: *Airline Business*, September 2003.

These facts are important because there are debates about the effects of scale on airline costs and revenues, and while the broad consensus is that while pure size is not important beyond a threshold, economies of scope and density (and on the revenue side, market presence) have relevance. This implies that within the overall network served by a carrier there may be significant differences in unit costs for individual routes or sub-markets.

Although economies of scale in a narrow economic sense may not be large, size may be important in other ways. Bigger airlines usually have greater reserves and easier access to finance that can be important during downturns in their markets, and scale can be of political importance. The extent to which this is important depends on temporal context and the institutional structure of the country. Evidence suggests that airlines' unit costs do not fall greatly as they expand. Strictly the evidence indicates that within any city pair markets there are rapidly declining costs of service, but that these eventually flatten out. There are then approximately constant returns to scale for airline systems that have reached the size of the major carriers. Savings come from attracting more traffic rather than expanding the network to cover additional origins/destinations.

Standard economic analysis, however, focuses on firms producing a single output. This does not adequately reflect the complexity of relationships in the aviation industry. Virtually all airlines produce a range of outputs by operating more than one service on any given city-pair route and providing a number of interconnected ones. Economies of scope occur when it is less costly for one airline to provide a range of services across a fixed network than for a number of airlines to provide them separately.

In terms of market entry, economies of scope imply that entry needs to be across a range of markets if the costs of the entrant are to match incumbents costs. Successful entry of no-frills, point-to-point airlines into European markets demonstrates that costs can in some cases be lower for such services than for larger networks.

Regulatory reforms have seen airlines seeking diversity of service, primarily via hub-and-spoke operations. These types of service involve focusing services around key airports. Airport hubs can take a variety of forms. Doganis and Dennis (1989) separate hubs into 'hourglass' and 'hinterland' hubs (Figure 3.5). The former is operated with flights from one region to points broadly in the opposite direction. A hinterland hub feeds short haul connecting traffic to long haul, often international, flights. The hourglass hub operation tends to use aircraft all of a similar size, whereas the hinterland hub has aircraft of mixed sizes.

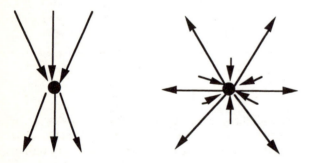

Figure 3.5 Hourglass hub (left) and hinterland hub (right)

With these forms of hubbing structure, flights are funneled in 'banks' into a number of large hubs where substantial numbers of passengers changed aircraft to complete their journeys. These banks involve the coordinated arrival of a large number of flights in a short space of time and then an equally coordinated departure of flights within a narrow time window. Larger hubs may well have up to seven or more such banks a day. Travel time would be longer for many people but fares fell and the range of potential flight combinations available to any particular destination expanded considerably.

Hubs have proved particularly important in the development of long haul air transportation and have served in many cases to circumvent the limitations of restrictive air service agreements. Such agreements, while facilitating flights between countries, have limited the ability of one country from collecting feeder traffic in another country or from distributing traffic to non-gateway destinations.

The growth of strategic alliances whereby airlines agree to serve as collectors and distributors of each others traffic for cooperative international services has provided a partial solution to this. In the Star Alliance, for example, the German airline Lufthansa may consolidate European passengers at Frankfurt (A in Figure 3.6) and then either it or its US partner United Airlines fly them to Washington (B). United then functions as the distributor to final US destinations. There is an increasing tendency over the denser long-haul markets such as the North Atlantic for competition between networks of these types each involving alliances of carriers rather than between single carriers offering point-to-point services.

Figure 3.6 Multiple 'bone-shaped' hub structure

The empirical evidence is not conclusive that this generates significant economies of scope. The main difficulties have been in isolating scope and scale effects from other aspects of airlines' cost functions. Further, a major rationale for hubbing comes from marketing advances (i.e., on the demand side). Providing a diverse range of services leads to market visibility and makes frequent flyer programs more attractive, thus enhancing customer loyalty. These attributes are features of network value or 'value of presence and utility to customer'. Smaller operators and new market entrants can, however, also enjoy some marketing benefits of diversity by forming alliances – British Midland vigorously pursued this approach in the past.

Economies from traffic density are often as important as economies of scope, although their effects are entwined. As more passengers travel, it becomes possible to use larger aircraft that are cheaper to operate per seat kilometer and offer more frequent services. Hub operations, by increasing city-pairs served, allows a carrier to utilize better its unsold seats inventory. Early empirical evidence from the US domestic market points to significant economies of density. This has been re enforced by more recent studies from several diverse markets.

Economies from operating a standard fleet of aircraft also exist, particularly when operating a point-to-point structure. Particularly, commonality of spare parts, maintenance procedures and flight crews can reduce unit-operating costs. This is exemplified in the US domestic market by the savings Southwest Airlines has achieved, partly through its reliance on Boeing 737s.

These economies are being exploited in short-haul markets, with few impediments to market entry. In some instances, airlines have made use of the large second-hand market to equip themselves from the outset with a standard fleet of aircraft. More recently, families of aircraft have increased the degree of commonality across aircraft types, a feature important for the full service network carriers.

Network economics

What exactly constitutes a hub-and-spoke structure is opaque. The US General Accounting Office generally assumes a concentrated hub to be an airport that was one of the 75 busiest in the nation in terms of enplanements and where one carrier accounts for at least 60% of enplanements or two carriers combined accounted for at least 85%. Airports falling into either category but were not in the 48 contiguous states are excluded, as are those in cities with more than one airport. But a more general rule has been a carrier feeding three or more banks of traffic daily through an airport from some 40 or more cities.

US major carriers generally dominate one or more large airports, with some secondary hubs designed to meet the demands of the domestic market. Flights are arranged in banks allowing passengers continuing on to be consolidated on outbound flights to further destinations. In the context of international flights, there are usually major hubs in each of the countries involved.

The European structure is different, with hub domination much less pronounced and with no genuine secondary hub structure. While economics is a factor, the hub-and-spoke structure has often been less a consequence of market forces and more the result of institutional arrangements. State-owned carriers have enjoyed extensive monopoly rights, both for domestic and international carriage, and have used this to create protected hubs. Relaxation of regulations within the EU area has seen new entry into domestic and short haul markets, with full service carriers' quasi monopolies being eroded.

Besides periodic congestion, banking of flights has been seen as posing few economic problems. But it can lead to periods of idle time when the number of banks per day is relatively small. Ground staff and other resources are left with little to do, and aircraft are used much less effectively.

Now it is recognized that while airlines may enjoy economies of presence from having large-scale hub-and-spoke operations, diseconomies be considered. Indeed, some, such as American Airlines, are smoothing oᴜ. ᴜᴜᴜᴋᴏ. Other air carriers are extremely successful in engaging in non-hub-and-spoke operations. They have either focused on taking radial traffic from some hubs to nodal points or have offered direct services between smaller cities. Several new EU start-up carriers, such as easyJet and Ryanair, can be categorized in this way, as can many of the regional carriers.

Long-haul international traffic can pose another challenge due to additional constraints such as differing time zones. Night curfews at airports and travelers' desires to arrive within certain time bands limit flexibility in operations. And the need for specific aircraft for long haul also affects a carrier's fleet composition. These factors may make it difficult for international carriers to provide hub type arrangements at both ends of a flight, although the size of aircraft would ideally mean that linked feeder services would be desirable. This is one of the reasons why increased cooperation exists between carriers of different countries; in effect, they seek to generate additional synergy effects from combining their respective hub activities.

Strategic airline alliances

In the past, airlines in Europe have formed alliances on a largely ad hoc basis to economize on aircraft maintenance. For example, in 1968, UTA, SAS, KLM and Swissair formed KSSU. Subsequent groupings, such as Atlas, were of a similar nature. Equally, the cooperation of scheduled airlines to develop computer reservation systems represents a form of alliance – the development of Amadeus in Europe was an example of this. The scope and nature of modern strategic airline alliances transcends these early efforts at cooperation.

Although *Airline Business* conveniently publishes an annual list, the exact number of airline alliances in the World at any one time is unclear. The *Airline Business* listing is voluntary and hence, invariably incomplete. But there are other problems brought about, not only because of the dynamic nature of the arrangements but because the term 'alliance' is generic, with no precise definition.

Alliances may involve equity holdings, but more often embody code sharing and cooperation in other ways. They may just involve a single route but the more important ones are across a large number of routes so as to strategically link networks. An oft quoted figure, however, is that about 54% of the world's global airline capacity is operated under broad multi-airline alliance networks.

All the major European carriers belong to a global alliance based around major US/European airline pairing. The situation in 2001 regarding the trans-Atlantic is seen in Table 3.3. This type of strategy dates back to the Global Excellence alliance formed by Swissair, Singapore International Airlines and Delta in 1989.

There were some major shifts in the structure of alliances in 2003. The Dutch carrier KLM began a process of joining much more closely with Air France (which paid $914 million to acquire it) and moved into the SkyTeam alliance (which embraced Delta, Korean Air, Alitalia, Aeromexico and CSA Czech as well as Air

France). This followed Northwest and Continental indicating a move to that alliance.

The result is that there are now three major global alliances that are taking an increasing share of the world's air transport market – SkyTeam with 19.5% of the world's global passengers, oneworld with 14.0%, and Star Alliance with 21.5%. The Star Alliance, with $22.8%, also has the largest share of global revenues. Within Europe, the Star Alliance in 2002 accounted for 25.1% of the traffic carried by the three strategic alliances, with oneworld taking 24.5% of RPK and SkyTeam 23.0%.

Airline alliances are formed for diverse reasons. Their members often hoped that they may improve technical economic efficiency that accompanies scale, help in the creation of optimal networks, create a unified marketing image, and provide a foundation for the coordination of services. But alliances also create market power and limit competition. They are also sometimes formed for the wrong reasons (Suen, 2002).

Table 3.3 Major North Atlantic alliances and largest carrier shares (2001)

Alliance	% seats on Europe-US	Airline shares	
oneworld	26.7%	British Airways	13.5%
		American Airlines	9.0%
Star Alliance	17.6%	United Airlines	8.5%
		Lufthansa	6.9%
Sky Team	17.2%	Delta	11.1%
		Air France	6.1%
KLM/NW	8.5%	Northwest	5.5%
		KLM	3.0%

Note: From 2004 Air France and KLM have merged and Northwest has joined the Sky Team.
Source: Association of European Airlines, 2001.

Objective evidence on the implications of alliances for airline efficiency is surprisingly scant. The emergence of strategic alliances involving transatlantic partners provides some indicators. These studies have primarily been of the early alliances but their findings are insightful. Studies involving government agencies instead of academics, have paid close attention to distribution implications since they involve carriers from the US and Europe. They have inevitably focused on the effects of existing alliances on their carriers and existing competitors. The findings of a number of these studies are seen in Table 3.4.

Table 3.4 Summary of major studies of transatlantic airline alliances

- Dresner et al (1995) looked at the Continental/SAS, Delta/Swissair and Northwest/KLM alliances up to 1989, examining how partners strengthened their international route structures, the extent to which an alliance would draw passengers from non-aligned competitors, and the effects of alliances on load factors. They found that those associated with the Northwest/KLM arrangement generated the greatest advantages in terms of increasing load factors, market share and realignment of strategies.
- Gellman Research Associates (1994) looked at the USAir/British Airways and the Northwest/KLM alliances for the first quarter of 1994 a time when the latter had matured but when the implications of the USAir/British Airways alliance were still evolving. They estimated the impact of alliances on the revenues, costs and profits of the carriers involved and other airlines serving common routes. Alliances generated benefits for the airlines involved and for passengers. USAir/British Airways and Northwest/KLM increased their market shares on code-sharing routes by 8% and 10% respectively. For British Airways, this represented $27.2 million of additional net revenue, and for USAir, $5.6 million. For Northwest, the alliance was estimated to benefit them by $16.1 million annually and KLM by $10.6 million.
- The US General Accounting Office (1995) looked at a number of international alliances. Those airlines participating in alliances benefited. Some gains came from generated traffic, but a significant amount of transfers came from non-alliance carriers. For example, the Northwest/KLM alliance cost Continental Airlines about $1 million in revenue in 1994. It pointed to a number of new (or reintroduced) international services resulting from alliances involving US carriers. Some alliances, by coordinating services of member airlines, offered more choice of carriers and routes.
- Brueckner (2003) examined the benefits generated by the Star Alliance. He found that the combined benefits of alliance membership, code sharing and antitrust immunity resulted in a 27% reduction in interlining fares. The benefits for passengers making non-stop trips were not considered, but an earlier study indicated that the alliance had no significant effect on these (Brueckner and Whalen, 2000). Additional analysis of the attempt by British Airways and American Airlines Alliance to gain anti-trust immunity suggests that interlining passengers would have enjoyed $40 per annum of benefits if it had not been stymied.

Overall, the empirical evidence seems to indicate that strategic alliances have certainly benefited the air carriers that participate in terms of enhancing their market share. But equally, evidence that this has been at the expense of the traveling public does not seem to emerge. The picture seems to be that the alliance airlines have both diverted traffic from competing carriers and through improved efficiency and ipso facto lower costs, have generated additional traffic on the routes involved.

Airline efficiency

Economic efficiency is normally measured in terms of cost per unit of output. Unit costs, however, often exceed the minimum when there are market imperfections, or when there are institutional structures that dim managerial incentive. Put simply, in a competitive market, a supplier only survives if it can keep its costs as low as its competitors and does not seek to exploit customers.

In the institutionalized bilateral European airline markets that preceded the Three Packages of reform there was limited incentive for airlines to keep their costs down, and in many cases competition was based upon what were perceived to be service quality factors (e.g., quality of meals, frequency of service, and timing of flights).

Many airlines also enjoyed significant state subsidies and other forms of support that can add to the potential for X-inefficiency. In measures to complement the Three Packages, the Union in 1994 set out principles and criteria for the assessment of state aid to airlines in the guidelines applicable to the subject under various European treaties.

Basically loans, capital, and guarantees can only be given by governments under the market economy principle – essentially a capital transaction would be considered state aid if an investor operating under normal market conditions would not be prepared to make an equivalent investment in the airline. Once for all phased restructuring finance was allowed to facilitate adjustments to the new market environment but this was under Commission monitoring. Airlines such as Air France, Alitalia, Olympic, and Iberia were amongst those that enjoyed such financing.

It is also certainly true that in naturally imperfect markets cost efficiencies are not the only determining factor for success in airline markets. Many other factors including service quality, good yield management systems, appropriate network coverage, marketing, and reliability can act as an offset for higher costs through greater revenue generation. Nevertheless, in the long term, in a market with free entry, those entrants leading to lower unit costs per unit of 'quality' offered will prevail.

Airline costs

The comparative cost structures of European airlines and other carriers provide some indication of how productive the former are, and how this productivity has changed over time. However, cost and productivity analysis is notoriously difficult to undertake. There can be important differences in market structures and in the institutional structures under which airlines operate. Consequently, findings should be treated with some care and are probably best seen as indicative of broad trends. There are a number of economic studies that have focused on international comparability of airline cost and production functions and some of these are summarized in Table 3.5.

Part of the reason for the higher European costs found in these studies is the lower productivity of labour. Labour costs are generally the second largest cost item for an airline. Europe's scheduled airlines, although improving, traditionally use labour less productively. In an early study, McGowan and Seabright (1989) found 8 US majors enjoying 1.6 million revenue passenger kilometres per employee compared to 1.1 for British Airways, the most efficient European carrier. An adjustment for different stage length finds British Airways labour 65% as productive as US majors. Later analysis by Oum and Yu (1995) confirms this, suggesting that British Airways and KLM were about 70% as efficient as US counterparts.

More recently, costs across 12 EU airlines were studied by Ng and Seabright (2001) were per pilot $151,200, for a cabin crew $48,500 and for other personnel $46,000 compared to $110,500, $30,700, $37,000 respectively for US counterparts. Million revenue passenger kilometres per EU employee were only 1.33 compared to 1.93 for compable US airlines.

Alamdari (1997), examining labour costs across EU carriers for the period 1991-95, found a fall of 38% per air transport kilometre with real wages per employee rising by 15%; the latter attributed to the outsourcing taking place that left higher skilled labour within the airlines. Ng and Seabright later estimated that labour costs could potentially fall by a further 15%-20% from their 1995 levels if EU airlines follow US patterns.

European air transport infrastructure

Air transportation relies heavily on the airport and air traffic control infrastructure in the markets that it serves. Much of this is provided independently of the airlines and there is a large public presence not only in the form of regulation but frequently involving state or local authority ownership. While there are some changes taking place with some airports and air traffic control systems being taken out of public ownership, market mechanisms are still largely usually absent in the provision and allocation of this infrastructure.

Airports

The majority of major airports in Europe are heavily congested. A succession of AEA surveys showed relatively high levels of departure delay in the late 1980s, with improvements generally in the early 1990s, up to 1994. After 1994, however, there was a gradual rise in departure delays as measured by the percentage of flights delayed by 15 minutes or more. The problems became particularly acute in 1999 but even after extensive efforts to improve the situation, 2000 was still the second worse year on record. The downturn in traffic from 2001 eased the situation.

Table 3.5 Studies of the relative overall efficiency of European airlines

- The UK Civil Aviation Authority (1983) found EU airlines costs were double US domestic trunk carriers.
- Starkie and Starrs (1984), comparing 27 non-US carriers with that of 21 US carriers for the five years to 1975 with the subsequent five years showed that while the growth in productivity of the latter continued at the pre-deregulation level, it declined by about 40% for non-US airlines.
- Sawers (1987) compared the costs of local European services with those of the US carriers Piedmont (which concentrated on business travel) and Southwest (concentrating on leisure travel) that offered services of similar length. In 1983 European carriers had costs per available seat mile of $0.142 compared with Piedmont's $0.090 and Southwest's $0.065.
- Barrett (1987) found that in 1984, the productivity of US airlines was 36% greater than their European counterparts in terms of traffic units per staff member.
- The relative international efficiency of US carriers was examined by Caves et al (1987). For the period 1970 to 1983 they found EU scheduled carriers to be less efficient than their US counterparts although pointing to the problems of making comparisons across very different route structures.
- Encaoua (1991) examined European airline costs and found some convergence between 1981 and 1986.
- Good et al (1993) compared the efficiency and productivity of four large European airlines with eight US carriers for the period 1976 to 1986. The same period was later used by the authors (Good et al, 1995) to conduct both programming and stochastic frontier analysis on eight European and eight US airlines, and found that the former would save about $4 billion a year (1986 dollars) if they became as efficient as US airlines.
- Distexhe and Perelman (1994) used programming methods to find reduced X-inefficiency amongst 33 airlines over the years 1973 to 1983.
- Oum and Yu (1995) conducted an international comparison of 23 airlines for the period 1986 to 1993. With the exception of the Japanese airlines, Asian carriers were found to have lower unit costs than European or American airlines. The period also saw important productivity improvements in the European carriers compared to their US counterparts; probably a catching up effect as these markets began to be deregulated.
- Ng and Seabright (2001), taking 12 European and 8 US carriers for the period 1982 to 1995 found that public ownership and market power resulted in European carriers being significantly less productive than their US counterparts. For 1990 to 1995 their costs would have been 26% lower if they had functioned as the US airlines, although the differential fell towards the end of the period as EU liberalization began to take effect.

Most European airports have traditionally been in one form of government ownership or another. On occasions this has been for strategic reasons. More often, the rationale is couched in terms suggesting that market failure exists (e.g., market power, local development effects, and environmental externalities) and government regulation or direct involvement is thus required. Recent years have seen moves to privatize and commercialize many of Europe's airports. This has often been done to improve their efficiency.

Juan (1995) suggests that the quality of service and investment commitments have significantly improved when the private sector has a major participation in management and ownership. He finds that airside charges have not varied so much in terms of their average level but the charges mechanisms have become more complex. Second, airside charges are now the subject of price-cap economic regulation. Finally, there has been intense development of high revenue yielding non-aeronautical commercial airport revenue.

About 25% of delays are weather related. The largest category of delays stem from airport and air traffic control difficulties. In 1996, air traffic flow management over Europe was centralized within Eurocontrol. While the move helped to alleviate delays in the worst affected sectors, it introduced delays in sectors which had previously operated with minimal delay.

Air traffic control

The European air traffic control system still remains fragmented with European ATC centres, national systems, hardware suppliers, operating systems and programming languages under the European Civil Aviation Conference (ECAC) organizational umbrella. Since 1999, however, the EU authorities have been moving forward with a Single European Sky inititative to improve the overall system (Commission of the European Union, 1999). The aim is to increase capacity and efficiency as well as eliminate delays and significantly reduce costs from the end of 2004.

The AEA has shown that en route charges and ground handling charges are the infrastructure costs that increased most on European routes following the implementation of the Third Package in January 1993. On average, landing charges remained unchanged since 1993, with ground handling increasing by 6.2% and en route charges increasing by 6.4% between 1993 and 1995. These costs vary significantly across the ECAC states as demonstrated in Figure 3.7 which shows the en route costs to overfly European states for a standard aircraft type and distance of 850km. Aircraft and passenger handling delays increased in significance in the most recent period, reflecting internal airline procedures as well as airport ground facilities and terminal conditions.

The capacity of the airports in Europe has tended to lag behind the overall increase in air travel. The air traffic congestion building up at the major European hubs has impacted on the nature of the networks that each now serves. Part of the change has come about because of the business models that the larger airlines are developing. In particular, London Heathrow and London Gatwick airports have seen a decline in the number of destinations that they serve as British Airways has focused on longer-haul routes. In contrast, there has been a significant increase in destinations served from Frankfurt/Main, Paris Charles de Gaulle and Schiphol as their major carriers have become more global in their orientation (Table 3.6).

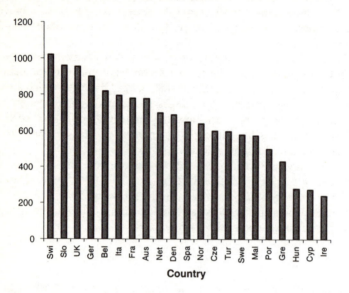

Source: Association of European Airlines, 1997a.

Figure 3.7 Costs to overfly Europe: Airbus A320 in 1997 for the average AEA airline's flight of 850km

This trend has also been exacerbated by the policy of national governments. The UK government, for example, has traditionally adopted a policy of fostering airline competition through a multi-airline strategy – for example between British Airways and Virgin Atlantic on North Atlantic routes. Hence slots have been allocated to competitor airlines to British Airways, the largest UK carrier, to provide competing services over existing spokes. Other European countries have tended to concentrate their long-haul services in the hands of a single flag carrier – Air France, Lufthansa, etc.

Assessments of EU air transport liberalization policy

The European Union's measures to reform European air transport are relatively new, and come within a much broader set of economic liberalization initiatives that extend well beyond aviation. In particular, the most important changes were part of the initiative in the late 1980s to create a Single European Market. Nevertheless, the specific set of institutional structures in which air transport services are provided within Europe have changed considerably within a comparatively short time.

Table 3.6 Number of destinations served from major European airports, 1990-2002

Year	Schiphol	Charles de Gaulle	Frankfurt/Main	London/Heathrow	London/Gatwick
1990	201	210	254	170	227
1991	198	210	250	142	212
1992	204	217	262	149	210
1993	210	219	271	147	212
1994	227	229	285	150	213
1996	231	254	317	165	209
1997	240	251	292	178	200
1998	239	251	291	179	198
1999	234	241	289	175	191
2000	238	250	292	180	189
2001	215	246	296	170	186
2002	205	220	296	140	191

Note: Includes all unique destinations served by direct flights and 2 leg flights.
Source: Parkinson and Sentence, 2002.

The change has not been neat and tidy, and was hardly achieved in a coordinated fashion, at least at the outset. There has been a path towards the major liberalization as initially Member States unilaterally deregulated domestic markets, then in pairs deregulated bilateral markets, and finally multilaterally relinquished regulatory responsibilities to the EU. The result has been a series of Packages of regulatory reform (Table 3.7 and Annex I) leaving market regulation within the EEA in a situation akin to US domestic markets.

Changes in overall EU competition and industry policy over the years have been designed to introduce more openness into markets. This has flowed in particular from the overall philosophy of the Single European Market initiative. Initial documents laying the foundations for reform brought up economic issues of contestability and considered the initiation of low cost point-to-point services by new entrants. In many ways these objectives have been met in the short term, although not always in the ways that might have been envisaged in the late 1980s when reform was initiated.

Table 3.7. The EU air transport 'Packages'

	1st Package From 1st January 1988 International scheduled passenger transport	2nd Package From 1st November 1990 International scheduled passenger transport	3rd Package From 1st January 1993 International scheduled passenger transport
Relevant Legislation	Regulation 3975/87 on the application of the competition rules to air transport, Regulation 3976/87 on the application of the Treaty to certain categories of agreements and concerted practices, Council Directive 87/601/EEC on air fares, and Council Decision 87/602/EEC on capacity sharing and market access.	Council Regulation 2342/90 on air fares, Council Regulation 2343/90 on market access, and Council Regulation 2344/90 on the application of the Treaty to certain categories of agreements and practices.	Council Regulation 2407/92 on licensing of air carriers, Council Regulation 2408/92 on market access, and Council Regulation 2409/92 on fares and rates.
Fares	Percent of fares approved: 'Discount' 66%-90% automatically; 'Deep Discount' 45%-65% automatically, and all others subject to double approval.	Percent of fares approved: 'Fully Flexible' 106% unless double disapproval, 'Normal Economy' 95%-105% automatically, 'Discount' 80%-94% automatically, 'Deep Discount' 30%-79% automatically, and all others subject to double approval.	Provisions made for the States and/or the Commission to intervene against excessive basic fares (in relation to long term fully allocated costs) and sustained downward development of fares.
Designation	Multiple designations by a State allowed if: 250,000 pass (1st year after integration), 200,000 pass or 1,200 rt flights (2nd year), and 180,000 pass or 1,000 rt flights (3rd year).	Multiple designations by a State allowed if: 140,000 passengers or 800 route flights (from January 1991) and 100,000 passengers or 600 route flights (from January 1992).	No longer applicable.
Capacity	Capacity shares between states: 45/55% from January 1988 and 40/60% from October 1989	Capacity shares of a State of up to 60%, and capacity can be increased by 7.5% points per year.	Unrestricted.
Route Access	3rd/4th freedom region to hub routes permitted, 5th freedom traffic allowed up to 30% of capacity, additional 5th rights for Irish and Portuguese, combination of points allowed, and some exemptions.	3rd/4th freedom between all airports, 5th freedom traffic allowed up to 50% of capacity, public service obligations and certain protection for new regional routes, a 3rd/4th freedom service can be matched by an airline from the other State, and scope for traffic distribution rules and restrictions related to congestion and environmental protection.	Full access to international and domestic routes within the EU (exemptions for Greek islands and Azores), cabotage unrestricted from April 1997 with restricted cabotage allowed for up to 50% of capacity until then, reformed public service obligations and some protection for new thin regional routes, and more scope for traffic distribution rules and restrictions related to congestion and environmental protection
Competition Rules	Ground exemption regarding; some capacity coordination, tariff consultation, slot allocation, common computer reservation systems, ground handling of aircraft, freight, passenger, and in-flight catering, and some sharing of pool revenues.	Ground exemption regarding; some capacity coordination, tariff consultation, slot allocation, common computer reservation systems, and round handling of aircraft, freight, passenger and in-flight catering.	Ground exemption regarding; some capacity coordination, tariff consultation, slot allocation, common computer reservation systems, and joint operation of new thin routes.
Licensing of Air Carriers	Not provided for in 1st and 2nd Packages.		Uniform conditions across EU; notion of EU ownership and control with small carriers subject to looser regulatory requirements.

Air carriers that have not had a tradition of market competition but have had to rethink their approach to service provision, as well as restructure their operations have also met them. These changes have proved more challenging than those that confronted the US domestic carriers after 1978 because of extensive public ownership of many flag carriers and the public utility ethos in management that goes with this. Nevertheless, the flag carriers have significantly modified the way they operate.

Assessing the implications of change are not easy, especially when they are phased rather than being a sudden one-off reform. Changes to EU air transport policy came at a time when there were also major structure changes taking place within the economies of Europe, and notably shifts away from manufacturing towards services industries. There was also increased trade more generally within Europe as the effects of the Single Market initiative were felt, and externally as global markets were beginning to emerge. Higher incomes were leading to new patterns of tourism that affected leisure travel markets. In these circumstances, defining a counterfactual against which actual events can be assessed is difficult.

Early statistical analysis by the UK Civil Aviation Authority (1994b) indicated that the reforms of the 1990s produced greater competition in terms of multiple airlines serving various market areas on domestic routes and international routes within the Union. The proportion of domestic round trip flights with two or more competitors rose from 26% to 36% between December 1992 and December 1994. Comparable figures for international flights were a rise from 19% to 25%.

Similarly, Schipper (1999) looking at 34 inter-city pair markets between 1988 and 1992 found on fully liberalized routes, the standard economy fare had fallen by 34% and the departure frequency had increased by 36%. Changes varied but countries such as Greece and Portugal considerably increased the number of competitive international services within the EU area. Many routes, however, where multiple services are simply not technically sustainable or institutional impediments still limited market entry remained monopolies.

The Commission of the European Communities (1996) in examining the impact of the Third Package subsequently found evidence of important consumer benefits. These included a rise in the number of routes flown within the EU from 490 to 520 between 1993 and 1995. Additionally, by 1995 some 30% of EU air routes were served by 2 operators and 6% of routes by 3 operators or more. Eighty new airlines had been created while only 60 disappeared, fares had fallen on routes where there are at least three operators.

Overall, after allowance was made for charter operations, 90%–95% of passengers on intra-Union routes were traveling on reduced fares, a caveat being that there were quite significant variations in the patterns of fares charged across routes. Mandel (1999) found that on routes with three or more competing airlines (about a third of the total in terms of passengers carried) fares fell.

There had been little change in fares on routes that remain monopolies or duopolies. The number of 5th freedom routes doubled to 30 between 1993 and 1996 although this type of operation remained a relatively small feature of the market and 7th freedoms were little used. Indeed, much of the new competition has been on domestic routes where routes operated by two or more carriers rose from 65 in

January 1963 to 114 in January 1996 with the largest expansions in France, Spain and Germany. The charter market has also continued to grow.

A UK Civil Aviation Authority (1998) study also looked at the evolving market. It found indications that the dominance of the major carriers declined somewhat with reform with regional carriers enjoying a larger market share. The only country where the full service carrier did not still control more than 70% of the market, however, was Ireland where Ryanair had 60%. Underlying this, much of the growth has been in short haul markets, with larger carriers strengthening their positions on inter-continental routes.

The study also looked at changes in fares between 1992 and 1997 for four markets (Amsterdam/London, Brussels/Rome, Madrid/Rome and Madrid/Milan) and found a mixed outcome. Fares had not fallen dramatically and, for some classes, especially business classes, there had been a rise. A more recent EU study of airline performance (BAE Systems, 2000) reveals that while promotional fares fell between 1992 and 2000 by 15% there were rises in business fares (by 45%) and economy fares (by 14%) in nominal terms (Figure 3.8).

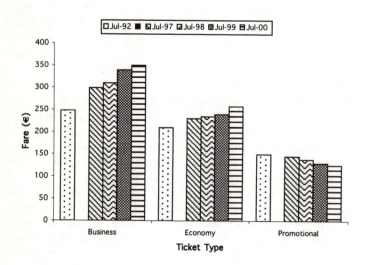

Source: BAE Systems (2000).

Figure 3.8 Weighted average fares within the EEA

There were regional differences in the patterns of fare changes. In particular, between 1999 and 2000 business class fares increased relatively more for northern EEA (up by 6.5%) routes than for southern routes where they fell by 2.1%. The converse was true for both economy and promotional fares. This may be seen as actions by the airlines to more accurately reflect local market conditions in their fare structures and to bring fares closer to costs.

These fare affects had been accompanied by a general increase in the number of scheduled carriers within the EEA over the business cycle – e.g., from 77 in 1992 to 140 in 2000. The largest number of new carriers was in Italy (from 6 to 19) although the UK and Scandinavia (each with 22 carriers) were the largest bases of registrations.

The adverse market conditions since September 2001 have led to what is possibly some short-term volatility in the market that extends beyond the more publicized demise of Sabena and Swissair. In terms of the scheduled market, from September 2001 to May 2002 17 airlines withdrew and there were 14 start-ups. In the French market the number of airlines fell from a peak of 26 in 1996 to 12 at the end of 2000 to 6 by mid-2002. There were also significant turnovers in Sweden and Greece.

The changes in the market environment within Europe have brought with them significant productivity improvements for the scheduled European carriers. This was alluded to earlier in terms of academic studies looking at converging rates of productivity growth in Europe and the US but is more simply illustrated in Figure 3.9 which provides a time trend of the labor productivity of the major European scheduled carriers.

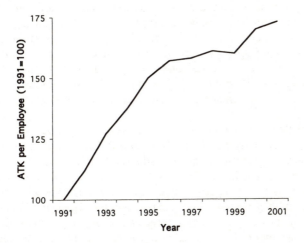

Source: Association of European Airlines.

Figure 3.9 Airline productivity in Europe

While the various Packages of reform have removed many economic controls within Europe, and action by Member states has reduced the restrictions inherent in many external air service agreements, some regulation remains and there are frequently efforts to modify them. In some cases the benefits of such measures (e.g. in terms of safety and security) often clearly outweigh any costs, but this is less obvious in other areas.

The challenges often relate as much to the de facto implementation of regulations as to the legal structures themselves. Many regulations involve high transaction costs for the airlines, as well as direct costs of implementation. They also often fall more on the larger scheduled carriers than smaller ones. In some cases there would seem to be justification for arguing that policy makers are passing the burden for policy formulation onto the airlines (e.g., in terms of defining practice governing conditions of carriage and passenger rights). While these may be seen as short term matters rather than strategic concerns, they pose burdens on an industry that operates with small profit margins.

Recent proposals from the Commission of the European Communities (2001), and largely approved by the Parliament, to mandate levels of compensation for passengers denied boarding because a planes is overbooked or for other reasons typifies this problem. The scheme finally proposed by a joint Parliamentary and Council of Ministers Committee raises the set level of compensation for cancelled flight from €150 for flights of less than 1,500 kilometers and €300 for those of greater length to €250 for flights below 1,500 kilometers, €400 for flights of 1,500 to 3,500 kilometers and €600 for longer flights. New rules were also established for overbooked flights.

There are good reasons for ensuring that those denied boarding are compensated. But in many cases they are passengers who are paying fares that only cover marginal cost and the proposed compensation level would therefore seem high. Additionally, the penalties entailed would lead to changes in the over booking practice of airlines (typically they are confronted with up to 25% no shows) that would lead to higher overall fares. The US system of essentially bidding up the compensation within a set of rules until sufficient volunteers have been found to equate supply and demand provides a benefit for both customers and airlines.

The central point is that in a market context there are choices to be made, and consumers have the option of covering themselves against delays in the insurance market, and the amount of coverage is an individual's decision. There may be arguments to support a minimal level of compensation, although it is not quite clear what its foundation is in a competitive market context, but there are serious moral hazard issues in having standard and penal structures. There may also be arguments for potential passengers having information about previous delay records so that they can make decisions about whether insurance is a sound individual decision.

A second area of short-term concern is that of war risk insurance. Full service airlines have covered their own insurance costs. The events of September 2001 changed this with the private insurance market basically withdrawing from providing this type of cover. Some respite was offered by the EU's member states' governments in the form of guarantees but there have been measures aimed at ending this (Commission of the European Communities, 2002).

There are sound reasons for airlines to take up many forms of insurance – e.g., there are safety issues and airlines have their particular responsibilities. The difficulty with war insurance has always been that it is a problem of uncertainty and not risk. [2] Hence normal actuarial principles do not apply. The tendency is, therefore, for private insurers to be over cautious and levy high premiums. There is always a danger in state interventions distorting markets, but there is also the problem that lack of appropriate state actions can damage markets.

---

[2] Coughlin et al (2002) provide some basic economics on matters of airline security.

Chapter 4

Is the Market Segmenting?

Introduction

The traditional market for intra-European commercial air services was integrated prior to the enactment of the Three Packages in the sense that the scheduled flag carriers provided a full menu of service types (and often have significant interests in air cargo transport) with charter carriers catering for a large volume of seasonal leisure travel. Some regional airlines filled in gaps by providing a range of local and specialized services. However, the market for passenger air transport in Europe now seems to be becoming much more segmented as potential consumers have the opportunity to reflect their preferences more clearly. The resultant reaction has been a greater segmentation of the supplying industry.

To some extent this pattern also seems to be a trend in air cargo operations. In the past a considerable amount of cargo was carried in belly holds of aircraft, but increasingly the role of specialized carriage is growing. These structural changes have involved a significant growth in the importance of integrated carriers to meet the needs of just-in-time production. This is a pattern that can be observed in many cargo markets and is certainly influenced by the growing interest in supply-chain management by industry in general.

Most of the rest of the focus here will be on the emerging passenger markets in the EU. But first there is the non-trival matter of how to define categories of airlines in a very rapidly changing world.

Categorizing airlines

Segmentation has effects on economics of the various categories of carrier. But to examine this it is first necessary to develop some categorization of airline types. Any form of categorization, and its detail, depends upon the purpose to which it is to be put. Here a fairly simple set of categories are examined to explore their dynamics and the level of segmentation that is emerging in Europe. It relies mainly on the types of service offered.

More detailed categorizations are possible, and indeed others have produced them, but they do give somewhat different pictures depending on the way that the cake is cut. It is possible to use a top-down approach making use of data on airline operating characteristics (e.g., RPK, ASK, revenue, etc.) to seek out statistical similarities. This offers a way of categorizing the position of airlines in terms of a selected scale.

The alternative is a bottom-up approach that is more intuitive but reflects the approach of carriers to the market. In this sense it has the potential of being more dynamic and possibly offering a more useful categorization for predictive exercises. It can also offer insight into the complexity of the industry's structure which is important for policy refinement. For example, ICF Consulting/Button (2003) by looking at airlines from a basic business model perspective to reflect behavioural similarities developed 6 categories of airline;

- Global carriers with 'first tier' hubs.
- Intercontinental carriers with 'second tier' hubs.
- Exclusively intercontinental carriers.
- Smaller 'national' carriers designed to link their country into the European network.
- Exclusively intra-European carriers with radial networks.
- Charters.

Horan (2002) adopts a somewhat different bottom-up approach by looking at the way airlines use hubs, and in particular the difficulties that some type of carriers have in covering costs when there is an increasing amount of low yield traffic. In this way his classification focuses on hub access and use rather than a more conventional business model approach.

At the other extreme very simple categorizations have been used. These often just entail separating out the legacy, full service carriers and the no-frill operators. The aim of such a simple dichotomization of the market is often to allow focus on a particular class of carrier. Recently, for example, there has been considerable interest in the future of no-frill carriers as their market share has risen, some have proven to be highly profitable, and as new markets in Europe are emerging as enlargement takes place (e.g., Binggeli and Pompeo, 2002; Mercer Management Consulting, 2002; European Cockpit Association Industry Sub Group, 2002).

The approach here is in sympathy with the business model approach but does not seek to go into immense detail because it is not part of a predictive exercise. It separates the European airlines into full services carriers (what are often called legacy carries in the US), non-scheduled carriers (essentially the charter airlines), no-frill carriers, and premium only services.

Full services carriers

The full service providers seek to meet a range of market needs through radial networks and multi-configured cabins. Price discrimation is at the center of the yield mangement that is widely deployed to generate the revenues to sustain this structure. Other carriers that are now taking segments of that market from the full service airlines affect the intregraty of the full service carrier's network adversely, affecting its cost structure and its revenue generating capacity. This is because margins are low and the ability of the full service network airline to cover costs is highly dependent on incomes from the totality of the spokes served. Losing traffic from one spoke can affect the viability of services connecting other spokes to the hub. These carriers are equally dependent on the mix of traffic they carry. Business traffic is particularly important because of the yield it can generate.

The shift to a much more competitive environment and the gradual privatization of the full service carriers has inevitably posed some problems. Some of these stem from on-going contractual agreements with labor and other suppliers that developed in the days of public ownership. Flag carriers were traditionally tied to a single airport hub. There is a lack of experience in the managing of large private sector network carriers, and the experiences from elsewhere only provide partial information. There is also a residual of political involvement in some of the airlines whereby they are still seen as vehicles of the state, although management and ownership are now operating in a commercial world.

Networks

The full service European carriers have provided network services through their domestic hubs. They normally have a domination of the hub although not to the same extent as the US hub carriers. This is mainly due to the history of international bilateral agreements that effectively prevented a carrier from getting more than 50% of slots. The traditional European full service carriers provide a range of different services (business/leisure, domestic/intra-Europe/Intercontinental, home market/ sixth freedom, etc.) and this entails operating a mixed fleet of aircraft types.

These carriers make use of hub-and spoke style operations to maximize economies of scope and density in a similar way to the large US airlines, but are more limited in what they can do because of the shorter hauls in Europe. They have significant infrastructures to handle sales, marketing and distribution across a highly complex network. Much of their revenue has come from the business market, largely because their hubs are at the major centers of commerce, their services are frequent, and because their service networks, usually involving strategic alliances, offer on-line global connections. They also provide services tailored to the perceived needs of the business traveler such as lounges, differentiated cabins, and flexible ticketing.

The US structure of hub operations developed fairly rapidly after deregulation and was forged mainly by market considerations. The much slower path of regulatory change in Europe combined with the continuing needs of the the wider international air service agreement structure has resulted in considerably less change after deregulation.

The constraints of a heavily congested system of airports and air traffic control also limit any dynamic change in the hub structure. This legacy effect coupled with the reality of present constraints, has produced what is often seen as a two tier structure of hub service providers – those based at a major hub and those at a smaller hub. The major hubs, and their related carriers, are normally seen as Heathrow (British Airways), Frankfurt (Lufthansa), Charles deGaulle (Air France) and Schipol (KLM). In aggregate they account for some 55% of the scheduled capacity of the larger airlines (Table 4.1).

Table 4.1 Flights by the three lead carriers at US and European airports

Airport	Carrier 1	Carrier 2	Carrier 3
Top 10 US airports ranked by passengers			
Atlanta	Delta 73.7%	AirTran 14.6%	American 2.3%
Chicago	United 47.0%	American 38.6%	Delta 2.2%
Los Angeles	United 30.8%	American 19%	Southwest 13.8%
Dallas/Fort Worth	American 63.2%	Delta 25.7%	United 1.6%
Denver	United 53.3%	Frontier 15.1%	Great Lakes 12.2%
Phoenix	America West 51.1%	Southwest 27.7%	United 3.6%
Las Vegas	Southwest 39.6%	America West 20.3%	United 8.7%
Houston	Continental 82.7%	America West 3.4%	Delta 3.0%
Minneapolis	Northwest 80.3%	American 3.6%	Delta 2.9%
Detroit	Northwest 79.4%	American 3.8%	Delta 2.8%
Top 10 European airports ranked by passengers			
London Heathrow	British Airways 41.6%	bmi 12.1%	Lufthansa 4.8%
Frankfurt	Lufthansa 59.4%	British Airways 3.6%	Austrian 2.9%
Paris Charles de Gaulle	Air France 56.6%	British Airways 5.15%	Lufthansa 4.9%
Amsterdam	KLM 52.2%	Transavia 5.5%	easyJet 4.3%
Madrid	Iberia 57.0%	Spanair 12.7%	Air Europa 7.1%
London Gatwick	British Airways 55.1%	easyJet 12.8%	flybe British European 5.6%
Rome	Alitalia 46.2%	Air One 10.0%	Meridiana 3.9%
Munich	Lufthansa 56.8%	Beutsche BA 6.6%	Air Dolomiti 6.5%
Paris Orly	Air France 64.2%	Iberia 8.2%	Air ittoral 3.6%
Barcelona	Iberia 48.5%	Spanair 9.4%	Air Europa 5.5%

Source: *Airline Business*, June 2003.

As seen in Table 4.1, this represents less domination of most airports than is generally found at major US hub airports. But each is also a major intercontinental hub as well as an intra-European hub unlike most of the large US airports where international activities are a small part of the total.

Smaller airports often have links to these major hubs through the activities of alliance partners or subsidiary carriers. There is also a trend towards airports, or their owners, forming their own alliances to, for example, benefit from the bulk buying of inputs as with BAA and Copenhagen.

Finances

The gross revenues of these full service carriers, and those of their US counterparts have not been meeting their long run costs across the trade cycle, and these carriers finding flows of revenue sluggish after the September 2001 tragedy. But it is the operators at second tier hubs that have been suffering the most difficulty (Horan, 2002). There have also been important changes in the passenger mix being carried by the full service airlines. In particular there is relatively less use made of business fares, with a very notable decline in the shorter haul markets (Figure 4.1). This has been a long-term trend and has inevitably contributed to the finanacial problems that some carriers have been experiencing in the short haul market.

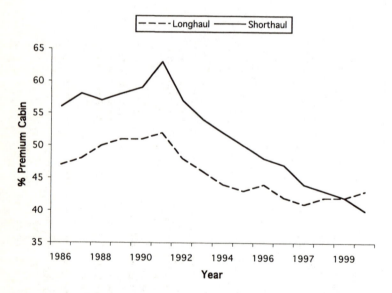

Source: British Airways.

Figure 4.1 The proportion of UK business passengers traveling business class

Recent events have led to restructuring of the activities of several of the hub carriers. Some of this can be seen as an acceleration of on-going strategy while in other cases there has been a strategic adjustment. A strengthening of global strategic alliances to allow more effective competitive placement has been a common strategy among the large airlines. There have been moves towards merger – for example Air France and KLM began a merger process in 2003. Sometimes in the past, however, such moves have been thwarted by the competition authorities. This effective consolidation has been part of a wider movement on the part of some full service carriers to shift resources into longer haul markets where significant economies of hub-and-spoke operations may be realized – British Airways has pehaps been the clearest example of this.

The full service carriers have also moved to confront the emergent no-frills carriers. This has been by modifying the structure of fares on their traditional intra-European networks – most notably offering more cheaper seats and relaxing some of the conditions attached to them – and in some cases establishing their own low cost carrier. A difficulty for the full service carriers is that unlike no-frills operations it is difficult for them to contract when confronted with new market conditions. A hub-and-spoke structure involves high levels of interconnectivity, and withdrawing any single service affects revenue flows throughout the rest of the network. Consequently marginal adjustments, for instance when a no-frills operation enters a market on a single spoke (or when there are cyclical reductions in aggregate demand), can be difficult to make.

Non-scheduled services

The passenger charter market in Europe is large – some 62 million passengers in 1999 doing 120 billion revenue passenger miles. This amounts to 21% of the intra-European market and 35% of the market if domestic and regional service are excluded. The European charter airlines are also often very large undertakings (Table 4.2).

The main function of charter airlines has been to provide leisure based services, normally as part of an inclusive tour, for traffic from the north of Europe to resorts in the south. They grew up as a means of circumventing restrictive international bilateral air service agreements. Although the advent of liberalization within the EU has removed most legal distinctions between the scheduled carriers and charter operators, non-scheduled services remain an integral part of the European air transport system.

Non-scheduled carriers have, because of their strong tie-in with the inclusive tour industry, more control over the balancing of supply and demand than do scheduled carriers. Bookings are made in advance (which also has a significant cash flow advantage) and their terms are very rigid which helps ensure high load factors (Figure 4.2). It is not unusual for inclusive tour operators to consolidate their packages at relatively late dates.

Table 4.2 Large European charter airline groups (2002 by passengers carried)

Airline operation	Country/region	Passenger (thousands)	RPK (millions)
TUI Group	Europe	18,180	49,927
Thomas Cook Group	Europe	14,400	38,959
MyTravel Group	Europe	9,840	30,567
Air Berlin	Germany	6,600	12,726
LTU International	Germany	5,700	16,100
Air 2000	UK	5,690	14,104
Monarch Airlines	UK	3,790	10,510
Aero Lloyd	Germany	3,500	7,500
Spanair	Spain	2,240	4,651
Transavia	Netherlands	2,050	4,943

Source; *Airline Business*, October 2003.

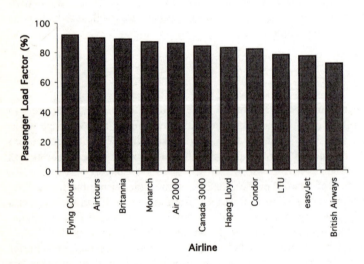

Source: UK Civil Aviation Authority.

Figure 4.2 Passenger load factors (1999)

The freer regulatory environment also now permits them to top-up their revenue with last minute seat only sales – it is estimated that about 20% of charter seats are 'seat only' sales. They operate from secondary airports and often at what would be off-peak times for scheduled airlines. They generally have more seats than comparable aircraft and fewer cabin crew. Despite this, many charter airlines have failed over the years.

As with much of the highly competitive tourist industry, charter airlines have always operated on extremely thin margins and have been highly vulnerable to shifts in overall economic conditions and to fluctuations in particular markets. Increased income, the development of new resort regions (that have the dual advantage of increasing demand and offering inter-seasonal peak destinations), and trends towards more but shorter vacations (which increases annual tourist trips) has kept the overall market buoyant. The full service carriers have provided more competition by developing non-scheduled services (carrying some 6.4 million passengers in 2001) and the no-frills carriers are capturing some leisure traffic from the charter carriers.

The nature of European leisure travel, however, and the widespread use of the inclusive package – which is entirely different to the US situation – indicate that this segment of the market will continue very much as before. There will inevitably be changes at the margin, and suppliers may come and go and merger take place, but the general structure is well established and seems fairly robust under all but very extreme scenarios. Indeed, there may be some counterattacks against the no-frills carriers – TUI, the largest package tour operator has given consideration to opening a no-frills airline.

No-frills services

An important change in Europe has been the growth in the number of no-frills carriers – low cost airlines to use US jargon. They began to emerge in the US during the days of economic regulation as intra-state carriers where federal fare and entry controls did not apply. These often trace their lineage back to Southwest Airlines and its services in pre 1978 Texas. Deregulation saw them grow in a variety of forms. While the business models of European carriers are often likened to the US low cost airlines, in many cases their business model is different (Cranfield University Air Transport Group, 2002). What is clear is that the expectation, certainly of financial analysts, is they are going to continue to flourish (e.g., Zonneveld, 2002; Merrill Lynch, 2002).

The carriers

While there are a rapidly growing number of no-frill carrier business models, two are of particular relevance. Ryanair (launched in 1995) is generally seen as the first no-frills European carrier. It offered low fare services on some 134 routes to 86 destinations at the end 2002 between mainly secondary airports, and breaks down the product so that elements other than the flight (e.g., food, and more than minimal baggage) are paid for separately. Bookings have risen to well over 80% through electronic means since a web site was established in 1999, telephone via a Dublin call center being the secondary means, and there are no interlinings. The airline does practice yield management although in a somewhat different way to the full service airlines. There are no class differences on aircraft and the fares charged tend to rise as the date of a departure approaches.[1] The full service carriers offer a wide range of fare/service options at any one time.

EasyJet has a somewhat different business model to Ryanair and does offer services to some of the larger European airports in competition with the full service carriers. Go, recently acquired by easyJet, offers a slightly higher quality of service and does serve larger airports, in part because it was initiated by British Airways as a potential way of competing with other no-frills operations.

The majority of no-frills traffic still involves UK links (just over 70% in 2003) but the number of carriers is growing very rapidly (a further 7 were added in 2001/2 to add to the 6 operating in 2000), and services within continental Europe are gradually taking a larger share.

The overall market share of no-frills carriers is relatively small – some 5.2% of the scheduled seat kilometers in the intra-EEA market in 2000. But the share is growing (from 0.4% in 1992). There has been dramatic growth from 2001 (Figure 4.3) with the six main carriers increasing the seats offered by 48.3% between the summers of 2001 and 2002, and reaching 1050 thousand seats by 2003. New carriers entered the market and some existing ones expanded; Ryanair offered over 550 thousand seats and easyJet over 500 thousand. The enlargement of the EU from 2004 (see Annex I) is likely to add to the impetus.

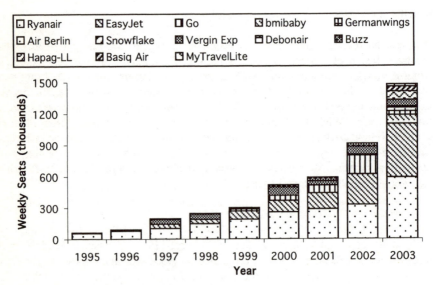

Source: Association of European Airlines.

Figure 4.3 Growth of no-frills carriers in Europe

---

[1] The situation would seem to be somewhat different when there are two competing no-frill carriers serving the same airports at origin and destination (Pitfield, 2004).

Whilst the primary no-frills markets have been between the UK and Ireland, and later between the British Isles and Continental Europe, and have generally been catering to new travelers, especially younger people, this is changing rapidly as the no-frills airlines are expanding services on Continental Europe.

Much of this growth has been the result of the no-frill carriers developing new base airports as well as expanding services from their existing bases. Ryanair, for example, has developed Frankfurt-Hahn and Charleroi as hubs and easyJet is developing Geneva and Amsterdam. Nevertheless, there were some 659 weekday departures on 312 routes in the summer of 2003 involving the UK, compared to 314 departures and 141 for Germany, 166 departures and 115 routes for Italy, and 152 departures and 78 routes for France.

Cost structure

While there is no single business model for the no-frills carriers, there are a number of generic features. No-frills carriers in general keep costs down in several ways. This involves entirely removing some elements of 'service' (or selling them separately) and in reducing the costs of others. This leads to a somewhat different business model for the no-frill operators. Table 4.3 provides a listing of some of the more common practices of no-frill carriers.

Since each carrier, no-frills and full service have their own unique features; accurate comparisons of cost structures are not possible. A number of broad estimates, however, have been made (e.g., Hansson et al, 2002; European Cockpit Association Industrial Sub-group, 2002), and Figure 4.4 provides a generalization of the relative importance of different cost items to full service carriers and to no-frills carriers.

No-frills airlines contain costs by striking hard bargains for their inputs (e.g., airports and aircraft) and making extensive use of their resources, and they cut items from their cost function wherever possible. In some cases, the no-frills carriers have transformed a cost item into a revenue item. This has been so in the case of Ryanair at some of its airports where it has received payments to provide services. This has recently posed legal issues where the EU has deemed that the payments form subsidies (e.g., at Strasbourg and Chareroi) but there is also some evidence that such payments, or favorable landing fees, may not prove a durable proposition as more airlines seek access to regional airports and, as a result, the airports develop more countervailing power (Francis et al, 2003).

In terms of more conventional revenue sources, much of no-frills airlines' traffic, as with Southwest in the US, is newly generated air traffic coming from road or rail, or from those who did not previously travel, but some does come from existing air routes long served by full service airlines. Binggeli and Pompeo (2002), for example, point to the rise in traffic between London and Barcelona from 600,000 passengers in 1996 to 1,500,000 in 2001 of which full-service carriers took 380,000 of the increase and no-frills carriers 520,000. The impact on yields, however, is less clear. Certainly if the US experience is being replicated, the yield of full service carriers was adversely affected.

Table 4.3 Features of European no-frills carriers

- European no-frill carriers often focus their operations on under-utilized secondary airports close to the key European metropolitan areas. For example, Ryanair, has focused its services around Paris (Beauvais), Brussels (Charleroi) and London (Stanstead and Luton airports). ∨
- There is one class of service with dense seat configurations and fewer cabin crew than full service carriers.
- Some, such as Ryanair, unbundle services and effectively charge extra for anything other than the flight.
- They negotiate with airports in the same manner to obtain the portfolio of services that they wish.
- There is no flexibility in the tickets that they sell.
- They keep their distribution costs low e.g., by not using global distribution systems and by often not giving travel agency commissions but rather, like easyJet, relying on the internet and call centers for their business.
- They differ from many of the early low cost carriers in the US (such as ValuJet) in that many use new aircraft that enable short terms drains on revenue for maintenance to be delayed until business has been built up.
- They provide a very limited range of point-to-point services. This allows aircraft to be turned around in 20 – 30 minutes compared to the 45 – 50 minutes required for the full service carriers whose aircraft arrive in banks to allow passenger interchange.
- They use their inputs intensively often making full use of the 900 hours cabin crew flying hours limit. No-frills carriers get about 11 daily hours flying out of each aircraft compared with about 9 for network carriers.
- Employees remuneration is often in part in stock option. This releases liquidity for expansion but in the longer term can pose labor relations problems if share prices do not grow as forecast.

Markets types

Because the no-frills carriers are a heterogeneous group it is difficult to generalize about the types of markets in which they compete. Unlike the US situation where the growth of low cost carriers was initially very slow, and more recently it has been contained by the FAA through tighter licensing after the crash of a ValuJet aircraft in Florida, the emergence of the European no-frills carriers has been relatively explosive. Consequently there has been a sort of learning by doing process as carriers have sought to see what European customers want and how the full service carriers will respond.

Carriers such as bmibaby, for example, were established to serve mainly traditional UK leisure routes and thus compete in markets where charter carriers have often provided a large part of the supply. Ryanair, although serving different airports, provides services that indirectly compete with full service carriers in some cases but charters in others. The airline has also has opened up new routes – such as

Stanstead to Ancona or Biaritz – where there is no direct competition. Certainly in the early years of liberalization in Europe there has been limited direct competition between the no-frills airlines.

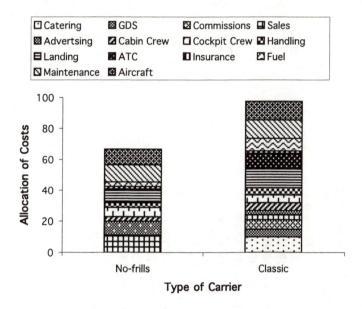

Figure 4.4   Typical cost differences on a flight stage between a no-frills carrier and a full service carrier

There are indications that the no-frills carriers may be moving more into traditional charter markets. Mytravel and Hapag-Lloyd Express (off-shoots of holiday companies) are opening up services from Birmingham and Cologne/Bonn respectively to a number of resort destinations, and existing carriers such as Loco Flights are expanding their services to new resort locations.

Nevertheless the base cost structure of the no-frills carriers (Figure 4.5) suggests that the lower costs of the established charter airlines gives them a considerable advantage. The differential is somewhat less when allowances are made for sector lengths because of the shorter sector distances flown by the no-frills airlines (Mason et al, 2000).

The former President of US's Southwest Airlines, Herb Kelleher, often described the carrier as a niche player that dovetails its services between those of the network-based majors. In a way the European no-frills carriers do this. But there are important effects of expectations of the low fares that these carriers can offer. Customers in other markets expect similar fares and it is not clear that these are consistent with financial viability. The full-service carriers have been slow to develop successful marketing strategies to counteract this.

Stability issues

There is a significant degree of instability in the no-frills market, in part because of the limited reserves of many of the players to meet downturns in the business cycle. This has resulted in carriers leaving the market (e.g., Debonair Airways, Euroscot, ColorAir, and AB Airlines) and mergers occurring (e.g., easyJet and Go in 2002). Further, whilst Ryanair earned an average annual operating margin of 23% between 1997 and 2001, the other no-frills carriers were losing 5% a year (Binggeli and Ponpeo, 2002).

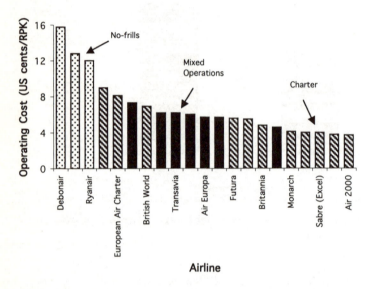

Source: Williams, 2001.

Figure 4.5 Operating costs of no-frills and charter airlines

There is an indication that these carriers may be encountering higher costs as they expand. For example, whilst Ryanair reduced its unit costs per air seat kilometer by about 40% from 1994 to 2001, easyJet (which often serves larger airports) saw its unit cost rise by 20% between 1999 and 2001. What experience is showing, both from the few years of no-frills operations in Europe and from other markets, is that many no-frills airlines fall by the wayside or fluctuate over time in the levels of service they offer (Barkin, et al, 1995).

Whether the no-frills airlines will expand in the future at the rate some of their statements suggest is open to some question. Certainly Airbus foresees orders by the no-frills for its wares declining after the current burst of activity aimed at moving into supplying no-frills airlines. The US experience has been that whilst Southwest has been extremely durable and successful there have been many other carriers that have entered and left the market when trying to emulate it. Commercial viability has not always been easy – the average operating profits for other US carriers was 2% between 1997 and 2001. Southwest expanded gradually, building up a network with only one major take-over. In contrast, there have been a number of mergers and take-overs in the European market.

In contrast, the expansion of the European no-frills carriers has been dramatic. Ryanair intends to expand its fleet, which consisted of 44 planes in 2002, and has placed orders for 100 new aircraft and options on 50 more by 2010 and easyJet intends to expand from 35 aircraft with orders for up to 136 new aircraft over the decade. But this implies they must increase the no-frills share of the European market to 14% by 2007. But this may not be easy. *Future*

There has already been take-over activity (e.g., of Go and Deutsche BA by easyJet and Buzz by Ryanair). This type of market consolidation offers some economies and reduces the potential for head-to-head competition between the no-frills airlines. In some cases such as the Ryanair acquisition of Buzz it also increased market power at a key airport, in Ryanair's case, Stanstead.

Competition between the no-frills carriers has been avoided to the extent so far that only 17 routes were being served by more than one no-frills carrier in mid-2002 whilst 111 had a single carrier. Larger airlines inevitably acquire higher fixed costs as their networks become more complex, and there is also the danger that, as with some US carriers, labor and other inputs demand more of the profits being earned. Mergers in the airline business have traditionally produced a plethora of managerial and industrial relations difficulties, and it is yet to be seen if this is also to be a feature of the European no-frills sector.

Much of the activity of low cost carriers has been in markets suited to their type of service and has been conducted at a time when countervailing factors have been weak. They have often entered airports on highly favorable terms (Frankfurt/Hahn costs Ryanair €4.25 per departing passenger and there is no landing fee; in contrast a 737 operator at Frankfurt/Main pays €13.00 per departing passenger and a landing fee of about €1.75.)

In some cases, regional airports are heavily subsidized from fees at hubs. Whether this is sustainable is an open question. Some airports seem to be using the lower fees as a lost leader to tie-in no-frills carriers, enabling fees to be raised later. In other cases, as demands at some secondary airports build up, market pressures will lead to higher fees.

The direct service nature of no-frills operations also has implications for the air traffic system. The hub-and-spoke system is economical on flights. Five cities can all be inter-linked by hubbing four services on one of them but it requires 10 flight segments to provide point-to-point service. The quality of service of European air traffic control is already poor, and has largely been designed for hub-and-spoke services. Additional services will stretch the system further and inevitably reduce the reliability of the services that they can offer.

The problem is exacerbated because the low margins characteristic of some no-frills carriers do not allow for the same size of back-up fleets as is the norm for full service carriers. Further, the turn-round times of no-frills equipment at airports (typically in the 20 to 25 minute range) does not allow any scope for catching up if a schedule fails. High load factors also means that rebooking a cancelled flight can pose problems. No-frills carriers do not simply reduce on-board service but also can reduce the quality of the reliability of the schedule offered.

The long-term economic issue is whether this type of service will prove attractive, particularly to business travelers, once economic recovery begins in Europe, and whether the structure of the European no-frills carriers is robust enough to ride out slow downs in their growth rate.

Regional carriers

In addition to the no-frills carriers there are numerous regional airlines in Europe that also often offer essentially no-frills services. The geographical limited coverage also means that even when there is a high quality of on-board service, they do not offer the range of destinations and service categories as the full services airlines. These are growing rapidly in number – the membership of the European Regions Airline Association has risen from 48 carriers in 1991 to 78 airlines in 2001. Some are also relatively large (Table 4.4).

Table 4.4 European regional airlines (over $200 million in revenue in 2002)

Airline	Country	Revenue ($ millions)	Operating margin	Net result ($millions)
BA CityExpress	UK	800	n/a	n/a
Eurowings	Germany	555	2.2%	5.0
SN Brussels	Belgium	442	-22.0%	-35.0
Air Nostrum	Spain	432	15.0%	46.4
Brit Air	France	365	1.2%	4.6
Régional	France	355	-7.6%	-43.7
KLM cityhopper	Netherlands	272	n/a	n/a
Widerøe's Flyveselskap	Norway	271	17.45	8.5
Air Litterol*	France	204	n/a	n/a

* Went bankrupt in 2004.

The regional airlines serve a variety of functions at the local level and on many thinner routes – e.g., smaller short haul markets and providing feed to larger carriers. They may be seen as complementary to the full service airlines they are often linked to and for which they often provide feed.

Premium only services

There are very few air services that are exclusively aimed at the business traveller. Most business travel is accomodated in a separate cabin on a multi-class aircraft. There have been attempts to develop such services in recent years (e.g. in the past by Trump in the US; whilst Lufthansa provides a single class service from Germany to the US and Ciao has been developing short haul business oriented services out of Italy) on the basis that business travelers are willing to pay a premium for high quality service.

The advent of long range private jets has also introduced a new competitive element into the business market that may pose a threat to the traditionally lucrative market of the full service carriers. How succesful the more recent innovations in this type of service will prove to be remains to be seen. If they do prove to be attractive, then there is the inherent danger that they will attract traffic from traditional network operations employing multi-class cabins. This will add to the problems of cost recovery through yield managment techniques for network carriers.

Segmentation by types of supplier

Markets can be segmented by the types of service provided, which is in effect what has been done here, but they can also be segmented by type of supplier. There is a strong correlation between the nature of the service offered and the type of airline providing it. But there is another way segmentation may be treated and this has more to do with the size of carriers and their linkages (notably various forms of alliances). The US market has, for example, a high degree of segmentation with major airlines and alliance partners providing national and international services, regional carriers, often allied to the majors, serving regional markets and providing feeder services to hubs[2], and low cost carriers filling a variety of niche markets.

The European market is much less structurally integrated than is the US. The former system of bilateral agreements within the EU led to national flag carriers without the presence of major pan-European airlines in the US sense. Regulatory change has gradually brought about some consolidation, most notably that involving Air France and KLM, but the structure of external EU agreements poses problems.

What is emerging in Europe, as indeed seems to be the picture globally, is that the various forms of network economy that become particularly dominant in the longer haul markets are leading to consolidation amongst the full service airlines in terms of strategic alliances and mergers. The difficulties of full merger may in some cases be the driving force behind this type of alliance but in other cases there are

---

[2]   These alliances between region airline and the full-service legacy carriers are not, however, always stable. In 2004, for example, in the US, Atlantic Coast Airlines, after a contractual dispute, ceased to provide feeder services for United Airlines in east coast markets and moved to become a low-cost carrier, Independence Air.

sound economic and managerial arguments for avoiding many of the problems that are often encountered in full mergers. These alliances are now based around the largest carriers (Lufthansa, Air France/KLM, and British Airways) and their global alliance partners. The no-frills carriers and charter operators are outside of this structure because of the simpler networks that they provide.

The degree to which further mergers, in the traditional sense, will take place and the extent to which carriers will enter and leave the market is impossible to foresee. In part it depends upon legal frameworks outside of the EU. But it does seem almost inevitable that natural market forces will move towards a situation more akin to that found in the US where the lack of international national boundaries has allowed mega carriers to emerge while at the same time permitting other airlines to fit into the wider industrial structure. The major difference in the longer term is likely to be influenced by airport capacity issues. The US market has far fewer constraints in this regard than has Europe.

Chapter 5

Is the Current Structure Sustainable?

Introduction

The concern here is with the ability of the European airline industry to generate sufficient revenue as a whole to enable optimal long-term investment to take place. This revenue may come directly from the fare box, but it can also be from financial markets when investors consider future net revenues will enable a reasonable return to be enjoyed.

The air transport industry is in many ways approaching maturity. While there is still the potential for significant long-term growth in many markets, the growth rate is slowing in others. The technology of the industry is relatively settled. The industry has gone through several technical revolutions but simple casual observation sees the commercial aircraft form has become standardized and airports are all very similar. This was not the case thirty or forty years ago. People also treat flying as an every day event rather than a special activity – they seldom dressed up for flights as they did in the 1950s and 1960s. Tickets are bought over the Internet in the same way as books.

Maturity, however, has traditionally posed particular problems for many businesses in terms of obtaining a continual flow of investment capital. With a continually expanding market there are successive pools of revenue to be drawn upon to meet capital costs. Financial markets offer more risk capital. Slower growth means that these costs of investment have to be borne by the existing pool of customers.

The ability of any industry to recover its full costs also depends upon the nature of the markets that it serves. Conventional economic theories of competition implicitly assume that input markets are also competitive. If this is not so then it is possible for a sector to be economically viable but for industries within it to be unsustainable. Effectively, the structure of the sector under such conditions means that those elements in the value chain with market powers leach away the net revenues of other elements by charging non-competitive, quasi-monopoly prices for their products and services. To be sustainable all elements in the value chain must earn enough to cover their long-run costs.

The cost recovery problem

For any industry to be economically sustainable it must be able to recover its full costs over the long term. It is quite common for industries, and individual firms within them, to lose money in the short term as demand fluctuates or there are sudden cost shocks. Perhaps most commonly it arise because of business cycle effects when there are temporary reductions in demand that are all part of economic cycles (Figure 5.1 provides some guidance to the correlation between business cycles and the use made of air transportation). But it also may be because of restructuring and the adoption of more efficient technologies, or it may be the result of bad management. The short-term losses, including any interest costs that may have been incurred to tide the industry through these times, must be minimized and recouped during more profitable periods.

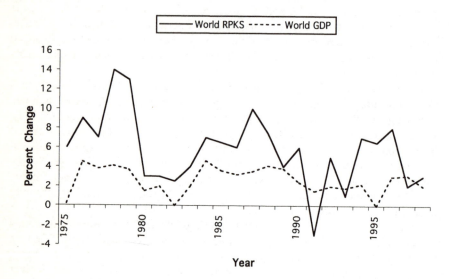

Figure 5.1 Air traffic and global GDP

The costs to be borne over this long period include capital. As with profits and losses, capital costs are not constant but vary over time; but if an industry earns less than the average for industry as a whole, investment will be attracted away from it. As seen earlier (see again Figure 3.3), the operating margin for airlines is volatile and there is also considerable variation between carriers that is not seen in aggregate data.

In addition to volatility, Figure 3.3 indicated that the operating margin for airlines as a group only periodically reaches the 2% - 4% range that might be seen, at best, as a minimum margin to allow recovery of full costs. The exact requirements depend on the methods of financing used, the perceived commercial risk associated with individual carriers, and potential earnings elsewhere. Even if it were only 1%, the picture is still one of inadequate margins. In fact it may be nearer 6% or even higher. But over the long term margins have consistently fallen short of even the lowest margins that have been put forward as viable.

The value chain concept

One way of setting the financial position of the European airlines in context is to view them as part of what Porter (1980) calls the 'value chain'. This involves the way in which undertakings create a competitive advantage. It is normally applied to a company but can equally well be conceptualized in the context of an industry or sector.

The value chain itself (Figure 5.2) is based on the process view of organizations, the idea of seeing a manufacturing or service organization as a system made up of subsystems each with inputs, transformation processeses and outputs. The inputs, transformation processes, and outputs involve the acquisition and consumption of resources – money, labor, materials, equipment, management, etc, and how the value chain activities are carried out determines costs and revenues.

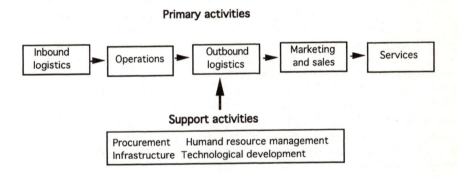

Figure 5.2 The notion of the value chain

The figure indicates how, in the context of a single company, goods are received from suppliers and stored until needed for production (inbound logistics), how these inputs are combined through some operations process to produce a product, and how outbound logistics then send these products along the outbound logistics supply chain of wholesalers and retailers to be marketed and sold. The services element reflects the need for after sales servicing. The nature and efficiency of the value chain determines final margins.

When dealing with linkages between companies, Porter talks about vertical linkages and argues that they are intrinsically the same as the linkages within the value chain of an individual company. The ability of the final link in the chain to earn an acceptable margin is influenced by what gains can be made on in the links before in the chain. This may or may not be a zero-sum game, although the latter is unlikely. Firms further up a vertical value chain may exercise market power and extract margins from those further down the line. This may in rare circumstances simply reflect transfers, but excessive extraction will not only reduce the margins at the end of the chain but also the overall value of the chain.

Evidence on the value chain

Airlines, to adopt Porter's terminology, can be seen as being at the end of a chain of vertical linkages that supply the ultimate air transport service. In contrast to the situation confronting the airlines, there is substantial evidence from a variety of sources that whilst airlines have trouble earning a reasonable return, their supplying industries do not (Figure 5.3).

Accountancy calculations by McKinsey conducted in 1998 and covering the years 1992 to 1996 indicate that the long-term rate of return on capital invested globally by airlines is about 6% – well below the long term rate for industry in general which is normally in the range of about 8% to 9.5%. Indeed, the European airline industry tends to earn even less than the 6% figure (about 4%). A subsequent study in *Airline Business* looking at the operating margins (a measure that avoids the need to place a value on capital) of various industries in the value chain on a global basis for 2000 to 2001 gives a remarkably similar profile. Airlines do extremely poorly when compared to actors further up the value chain.

To paint a specific European picture, an examination of operating margins for the years 1999, 2000, and 2001 was performed. These reveal a similar general pattern to the McKinsey and *Airline Business* studies. The data show not only that the full service airlines have financially performed poorly since the onset of the post September 2001 problems, but also that their net revenues were generally lower than most other elements in the value chain before that. In fact the picture painted by McKinsey, and the more recent repaintings are remarkably similar – the results are very robust.

There are particular problems in conducting this type of work and making comparisons across studies. They are inevitably incomplete because many of the players in the value chain are entirely or partly publicly owned, with different accountancy conventions, and are often missed.

Many of the private companies concerned are conglomerates, often with extensive international interest, and the net returns to their air transport activities are often not transparent from accounts. Aggregation across companies has the problem of deciding on appropriate weights to use. Financial years also differ between companies, largely because of variations in the laws associated with places of registry.

There are also frequent transmogrifications of companies. There has been consolidation in the airport handling and in the catering sectors. In recent years Penauille Polyservices has taken over Servisair and has a large interest in Globegrond. Candair (a financial services company with net profit margins of 67.3%, 62.8% and 43.5% for 1999-2001) has assimilated Swissport. In addition, GE Capital Aviation Services, a major aircraft leasing company has been subjected to restructuring within General Electric – International Lease Finance Corporation, is already part of AIG. Airbus has been combined with military activities in the European Aeronautic Defense and Space Company.

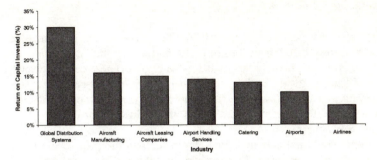

McKinsey calculations of return on capital invested (1992-1996)

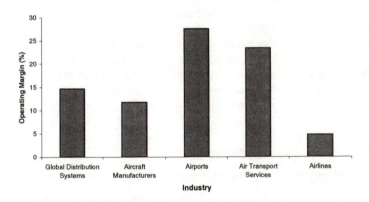

Airline Business calculations of operating profits (2000-2001)

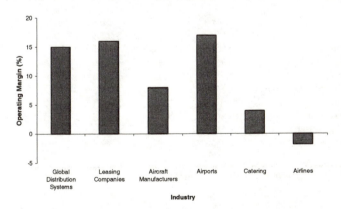

Operating profits for Europe from company accounts and AEA data (1999-2001)

Figure 5.3 Economic margins of industries within the air transport sector

Market power in the value chain

All these changes, however, would suggest that the market power of companies up-stream have consolidated their position. In particular, the involvement of financial service providers in activities up the value chain offers greater protection during crises. What this means is that the numbers in Figure 5.3 seem plausible, indeed conservative in some cases, are remarkably robust across approaches, and conform to what would be anticipated given the industrial structures of the elements in the value chain.

Of course one would not anticipate that all elements in the value chain would have identical operating margins. The methods used for finance matter, as do factors such as risk and the longevity of capital. But the difference seen in the figure would seem to exceed such differences.

The high returns enjoyed by many airport authorities are not surprising (Table 5.1 offers more details of the economic performance of Europe's top 10 airport operators by revenue). The global picture in 2002 shows all but Chicago, Detroit, and Hawaii of the world's largest airport groupings made profits; all major European airport groups were profitable. Although in some cities or regions there are a number of airports, in most there is only one or if there are more then they are congested. In either case there is de facto market power.

Table 5.1 Operating margins at European airports

Airport group	Operating margin (2001)	Operating margin (2002)
BAA plc (UK)	29.8%	30.6%
Fraport (Germany)	18.0%	15.8%
Aèroport de Paris (France)	6.0%	9.2%
Schiphol Group (Netherlands)	31.7%	32.0%
Luftartsverket (Sweden)	3.7%	9.1%
Flughafen München GmbH (Germany)	11.8%	3.7%
Avinor (Norway)	22.9%	17.1%
Aeroporti di Roma Spa (Italy)	16.8%	21.2%
SEA Aeroporti di Milano (Italy)	11.5%	10.4%
Manchester Airport Group (UK)	19.2%	19.3%

Source: Derived from *Airline Business*, December 2003.

The market positions of airports are often strengthened if international services outside of Europe are involved (Association of European Airlines, 1997a). Many airports in Europe are still publicly owned or, where privatized, are the subject of regulation, and have the ability to earn a return approximating the market rate. The hidden issue here, and one inherent in several elements of the value chain, is that such a system has limited incentive for the providers to act efficiently. A study by Pels (2000), for example, finds that European airports vary in their efficiency but overall there is a high level of inefficiency. 'Gold-plating' is one particular problem with airports often offering services that airlines do not need, or offering services at a level of quality beyond the needs of carriers.

There are only two major manufacturers of airframes for large aircraft (Boeing and Airbus) and only three manufacturers of large jet aero-engines. The number of regional jet manufacturers is only slightly larger. The main manufacturers of airframes have the advantage of being (or becoming) involved in military supply. This offers them a significant buffer to the adverse effects of downturns in the business cycle suggesting that their returns may be lower than normal market rates to attract necessary capital.

The EU has made concerted efforts to remove the monopoly power involved in ground handling by stipulating the number of handlers required at different size airports. Nevertheless, as seen in Table 5.2, there are a limited number of large international ground handling players that effectively operate in a world of local oligopolies.

Table 5.2 Major European based ground handling suppliers (2002)

Supplier	Base	Revenues ($ millions)	Stations
Serviceair/Global Ground	France	976	170
Swissport	Switzerland	648	150
Frankfurt Airport Ground Services	Germany	546	25
Worldwide Flight Services	France	512	100
Menzies	UK	357	91
Aviance UK	UK	319	17
AviaPartner	Belgium	249	32

Distribution systems are gradually changing as airlines begin to re-internalize the process (e.g., developing their own systems such as Opodo) but there are only four global distribution systems. These tend to be profitable and, indeed, several of these have made record profits in recent years (Table 5.3). This has to some extent been at the expense of traditional travel agents in some markets – travel agents like airlines generally supplying services in a relatively competitive environment.

There are seven aircraft leasing companies of any size but two of these – GEEM and ILFC – have about 45% of the global market. They are parts of much larger conglomerates – General Electric and AIG respectively – giving them considerable financial support in negotiations and a buffer through economic downturns. While the EU has attempted to introduce a degree of competition into airport handling, these measures have not been fully successful and at many airports a monopoly remains. Previous AEA studies have shown these charges to have risen significantly over time. In terms of catering there are two main players that control about 45% of the market with the largest seven controlling about 70% of the market.

Table 5.3 The main GDS providers, 2003

Company	Revenues ($ millions)	Operating margin	Operating result ($ millions)
Amadeus	2,195	16.6%	365
Cedent Group*	10,034	15.3%	1,534
Sabre	2,045	8.1%	166
Worldspan	929	9.7%	90

* This represents the company's total travel services, some 55% of its overall business.

The breakdowns of elements in Figure 5.3 miss an important component of airline costs – labor costs. Some 30% of labor employed by AEA airlines were flying staff in 2000 and of these 8% were pilots. The pilots, however, accounted for 22% of labor costs. A study by the AEA (1997b) found that this amounted to about 8% of all operating costs in 1996. Pilots are also limited in how much they may fly while changes in contract arrangements affect costs per hour flown. Consequently an increase in pilots' pay or work practices can have a significant effect on an airlines cost structure.

The pilots are highly unionized and unions represent a powerful monopoly bargaining unit. The methods of pay bargaining deployed amongst most of the European carriers leaves the airlines at a competitive disadvantage. They lose revenue rapidly when there is an industrial dispute because competitors take up the business, and there are adverse long-term effects on their market share. The unions are monopoly suppliers and hence there is no competitive labor market to restrain their actions. In particular, and unlike airframe manufacturers for example, their inputs are usually required immediately to allow any airline to expand its supply of service. This gives the unions ability to negotiate high wages during 'up' phases of business cycle – of which the 2000 pay increases at United Airlines is the oft cited extreme – but seniority clauses and, in the US, scoping clauses, make it difficult to adjust labor costs rapidly in downturns.

The table also misses the degree of secondary competition that affects each industry. European air transport faces strong competition from high-speed rail services that, with the exception of the Paris-Lyon TGV, receive direct subsidies. In some cases rail, particularly when linked directly into an airport, can provide useful feeder services to scheduled airline operations especially if it frees up airport capacity for more efficient airline activities. But when longer routes are concerned, subsidized high-speed rail is a significant source of competition.

In some cases, and especially where there are economies of scale in large-scale production, there are clear advantages in having a single supplier that can make use of scale economies. Indeed, hub-and-spoke operations in a sense require a large, quasi-monopoly airport to be successful. In other cases airlines may prefer to have an established link with a large supplier to reduce transaction costs as is often the case in labor negotiations. In all cases the issue is less the existence of the market power than whether that power is exercised. A supplier in the value chain, however, has an interest in maximizing its own long term profits and this is generally inconsistent with selling further down the chain at prices that do not yield some degree of monopoly profit to itself.

Is air transport's problem unique?

But does this lead to unsustainable situations for airlines that are some way down the air transport value chain? The inherent problem, and one that is found in other sectors, is that liberalization of one element has not been set in the context of other elements in the production chain. The notion that competition is efficient is premised on the fact that other sectors providing inputs or buying the product are also operating in competitive environments.

Perhaps the most dramatic recent example of the types of problem that can arise from differences in the structure of regulation in a supply chain is the energy crisis in California (Kahn, 2001). The overall situation was complex, but a key factor leading to the crisis was that the retail market was regulated but the wholesale market and the transmission grid, which was owned by incumbent monopolies, was not. The result was that the higher costs of energy that came about from fuel prices rises and restricted generating capacity could not be passed onto consumers and this produced a crisis in supply.

Telecommunications would seem to have similar problems. But the industry has only been deregulated recently and the overall projections during the Internet boom that it would grow by 100% per annum attracted vast capital inflows and led to a large construction program. The incumbent suppliers in Europe have been accused of exercising market power over those further down the value chain by being slow to open up their local networks. In this regard, both France Telecom and Deutsche Telecom have been investigated by the EU for their practices.

In the US, the former 'Baby Bells' have ownership of the last mile of the network that allows them to control access and their revenues despite legal requirements to permit open access. In the gas sector, the pipeline network is supposed to be open at regulated tariffs to gas merchants but cases are emerging of access being restricted.

But it is the nature of the distortions to the value chain that affect the ultimate outcomes. In the California energy case, it was final prices that were regulated preventing costs being passed on to consumers. This led to a serious and almost immediate crisis. This is not necessarily the outcome if the distortions are further up the value chain. In these circumstances, the final link in the chain can absorb part of the adverse effects of the distortion and pass these onto its customers. The cost is to the final supplier and the customer, generally in higher prices.

In general, these would seem to be the circumstances applicable to Europe's scheduled airlines. Their inputs are priced above the competitive level but this does not in itself lead to a market collapse. Rather it leads to a smaller than optimal output. Airlines essentially are forced to adjust their output downward to reflect the excessively high cost of their inputs (airframes, airport services, GDSs, etc.) or be unable to compete in what has become a highly competitive market. This not only has implications on the airlines' bottom lines, but also has adverse social implications as potential travelers are deterred from flying. But it should not mean that a normal profit cannot be earned by airlines. They should adjust their capacity to reflect the value chain effects.

Removing imperfections in the supply chain is certainly vital to ensuring that the full potential market can be supplied, but of itself it would not seem to be the underlying cause of the long-term financial problems of scheduled airlines.

The problem of full cost recovery

As we have seen, the airline industry has a combination of features that make it unique, and thus some of the problems it has in cost recovery are specific to it. But there are also many features that it shares with a range of other industries. It is helpful to separate the types of cost recovery issues that are a function of its peculiarities from those that are of a more general kind.

Generic issues of scheduled airline operations

Economic theory tells us that when there are no fixed costs then bargaining between suppliers and customers will ensure that prices are kept to a minimum consistent with the suppliers recovering all costs in the long-term. Since there are no fixed costs the marginal cost of meeting the customers' demand will be the entire costs of production. Once fixed costs are introduced in the situation when there is competitive supply then bargaining will push prices down to marginal cost, but they will be insufficient for capital cost recovery.[1]

In the case of scheduled air services, the airline is committed to tying up capital in each flight but then has to compete with other carriers for customers to fill seats. In their quest for business each carrier will quite rationally reduce prices because providing a passenger covers the marginal cost it is worth providing a seat. This is because a seat is in effect a fixed cost perishable product. There is a continual incentive to fill seats and fly underutilized capacity for fear that putting it (i.e., underused aircraft) on the market will add to competitive pressure. The fact that

---

[1]   This is not some obtuse part of economic theory. The Oxford economist Francis Edgeworth considered it in some detail as long ago as the 1890s, and subsequently the Nobel Prize winning economist Ronald Coase developed the topic further. The issue of stability is also considered in the context of contestability theory (Baumol et al, 1982).

full costs are not recovered and that ultimately the airline will withdraw the service or go out of business is known as the 'empty core problem'. In the long term, as potential investors become aware of this problem, they will cease to invest optimal amounts in the industry.

A simple explanation of the underlying problem is offered by Coy (2002) 'Think about why increasing-returns businesses are unstable. It costs a lot of money for them to gear up, but once they do, each incremental unit of output is extremely cheap to produce. Take airlines. Once they fill their planes enough to cover fixed costs, each additional passenger is enormously profitable. Trouble is, those profits invariably entice new entrants. Capacity increases. Fare wars begin. In desperation, airlines cut fares until some passengers are paying barely more than the cost of their meals. No longer earning enough to cover their fixed costs, airlines must merge or go bankrupt. Capacity falls, fares rise, profits increase, and the cycle begins again'.[2] In the long term this leads to sub-optimally low levels of investment despite excess capacity often being a short term problem during peaks in the cycle.

There are several conditions when there may be no core and, hence, a market may not be sustainable. These occur when there are relatively large fixed costs, avoidable (set-up) costs, indivisibility, or network effects, and severe fluctuations in demand, where some suppliers enjoy a degree of institutional or financial protection, and when there are significant variations in the costs of suppliers (Telser, 1978). In practice many public utilities, transport industries, and some manufacturing industries seem to have cost conditions in which a stable, efficient equilibrium is possible only by means of a suitable restructuring of the competitive environment.

This type of situation should be distinguished from some of the arguments for subsidies and market entry control that appeared in the past to protect declining industries (Kahn, 1988). This may have represented a core problem but it was of a different type. The argument was that destructive competition can emerge when fixed costs are a large part of overall costs and when there are long periods of

---

[2] Suppose that an economy has three individuals B1, B2 and B3 who wish to fly between two points. B1 is prepared to pay $100, B2 is prepared to pay $80 and B3 is prepared to pay $75 for the trip. There is one aircraft owned by A1 that can carry up to two passengers. The plane can make the trip at a cost of $140 whether it carries zero, one or two passengers. All of the sustainable outcomes have the property that B1 and B2 travel on the plane, whilst B3 fails to make a contract. The outcomes can differ in the rate that B1 and B2 pay for the journey. For example, the allocation in which B1 and B2 both have a flight and pay a rate of $75 is in the so-called 'core'. In all the outcomes in the core, airline A1 makes positive profits of at least $10. If the carrier lowered the rate further, below $75, the third individual, B3 would also wish to fly, potentially disturbing the coalition of A1, B1 and B2. There would be an excess demand at this rate and hence the core would be empty. If another identical aircraft, owned by A2, also with a maximum capacity of two and the same costs as A1, enters the market there would also be no core. Because the two aircraft are identical, and because equilibrium requires their owners to receive the same profit there is not enough demand to sustain both planes. They would compete prices down until one is driven from the market. The term destructive competition is sometimes used to describe this condition, although it also has other meanings.

excess capacity. In these conditions, the industry is 'unconcentrated' in that consumers are too few in number relative to the total size of the market to perceive and to act on the basis of their joint interest to avoid the competition that drives price down to marginal cost.

Historically the railways used this type of argument when confronted with lower cost road competition. Breyer (1984) has tended to discount the practical importance of this type of problem in declining industries, seeing it as a short-term structural issue. In the case of an empty core, the problem may have nothing to do with insufficient demand but rather the inability of the supplier to structure prices and services to tap into demand.

The potential relevance of this approach to the scheduled airline industry has been appreciated for some time but largely ignored in policy. The complexity of the underlying economic model has not helped in communication. It also runs counter to some of the more traditional views of competition policy where there can never be too much competition.

The analytical framework attracted attention in the US in the economic downturn of the early 1990s (Smith, 1995). It was rejected at the time by the airlines and by policy makers who believed that massive cuts in capacity was the solution to heavy financial losses being suffered by the scheduled carriers. They would bring the airline industry back to equilibrium. The period following this, which saw record economic growth in the US, did see profits being earned but even at the time of record income into the industry, the hub carriers only managed a net marginal profit of 2.9%. Subsequently from 2000, losses began to emerge again. While there is a need to keep costs down, and to adjust capacity to confront business cycle effects, which themselves may be difficult, the robust market for the services of several large low cost carriers over this period in the US would not indicate chronic excess capacity.

The trend towards market segmentation discussed previously does not necessarily help in this type of situation if each segment involves offering some form of scheduled service.[3] Advanced bookings and vertical integration remove the problem for charters. But the type of problem that full service carriers suffer is also seen in no-frills markets. While Southwest in the US is held up as an example of how successful a no-frills airline may be over time, carriers such as Reno, Kiwi, Midway, National etc. have entered the market and left in recent years.

The previous chapter highlighted comparable cases in the EU. They were not part of the instability of the initial shift to deregulation but rather participated when markets had been in place for some considerable time. Segmentation indeed may

---

[3] There have been arguments advanced that in network industries, such as air transport, stability can exist when there are a limited number of suppliers if some can earn relatively high returns on some parts of their networks to cross subsidize other parts whilst other carriers make their return on other segments. Mutual interest deters destructive competition with each supplier enjoying a degree of market power in segments of the system. The advent of point-to-point services can be seen to have disrupted this, although the evidence to date is mainly North American (Dresner, 2004).

make the situation worse if it removes the ability of full service carriers to price discriminate.

This is not a problem peculiar to airlines – although network industries do seem to have particular difficulties. The concern here is with the degree to which it poses a major difficulty to scheduled airlines and to examine the studies of the problem that may shed light on the scheduled airline case. However, to directly test for the existence of an empty core is not possible. If the core is empty then there is no equilibrium market-clearing price to use as a benchmark (as the competitive price is used when looking at market power). Consequently, while it is often possible to isolate factors that can lead to monopoly exploitation, and indeed put some values on them, this is not possible when there is a problem of excess in competition. As a result, assessments of whether empty cores exist have used indirect approaches (Table 5.4).

Table 5.4 Examples of studies of market instability

- Bittlingmayer (1982) in an historical study found evidence of empty core problems in US cast-iron pipe manufacturing.
- Bittlingmayer (1985) provided an explanation of many of the mergers that have taken place in terms not of market exploitation but of companies seeking to recover long-costs of capital.
- Telser (1985) looked at cooperation amongst lawyers on the basis that 'competition may require some cooperation in order to obtain efficiency' because of uncertainty in demand and the existence of fixed costs.
- Sjostrom's (1989, 1993) looked at shipping cartels (conferences) to seek consistency with an empty core. He looked for legal restrictions on entry (making the core theory for collusion less likely) or severe temporal variations in demand and costs (making the core theory more likely). Looking at the US he found that the theory of the core is supported.
- Pirrong (1992) put emphasis on exploring costs. He looked at cost, demand and market organization in scheduled shipping. The findings are consistent with Sjostrom's and suggest collusion and coalitions serve to ensure stability and avoid competitive chaos.
- Button (1996) looked at the changing air transport market within the EU. The analysis found that variables representing legal restrictions and market stability had signs associated with them consistent with possible empty core conditions. Evidence of the structure of the industry in the early 1990s supported this view.

Specific European problems

The fixed cost problem appears to be generic to all scheduled operations. This is overlaid in the European context by a number of institutional factors. The issue of the value chain has been discussed above in terms of its implications for overall social and economic efficiency of air transport. But it can also contribute to market instability if the higher costs further up the chain affect some carriers more than others (e.g., if the market power of catering suppliers is stronger at some airports than others). Equally, if airlines are confronted by different national laws regarding labor practices this violates the normal market conditions for stability.

Variations in bankruptcy laws have a similar distortion effect. The European airline industry, because of labor laws, is also the subject of very considerable shifts in demands for its output relative to its ability to adjust supply. These are all features that, while partly institutional, contribute to instability problems and the inability of the European airline industry to recover full costs.

The competition policy of the European Union compounds the problem in that it seeks to maximize competition without taking cognizance of the industry's need to manage capacity. While alliances and mergers can concentrate market power, they can also be mechanisms for introducing stability into an industry and offer customers a better-integrated structure of services. Many judgments are, however, based purely on criteria more appropriate for assessing natural monopolies. In particular, markets are often defined very narrowly. Institutionally the lack of a single external authority of air service agreements acts, because of the bilateral nature of the current regime, as a constraint on consolidation of carriers through mergers. This has been seen in the past when British Airways and KLM sought to link up.

Methods of capital cost recovery

The underlying cost recovery problem is far from new and is applicable to many modern industries. The financial problems of the telecommunications sector, for example, seem superficially to be of this ilk. But here the picture is muddied by its relatively recent deregulation, the rapid technical changes that are taking place, questionable accounting practices in some cases, and the nature of the licensing systems in place in Europe. Certainly the industry is losing money – total debts of the telecoms amount to over $1 (US) trillion – and there are significant defaults – globally amounting to over $30 billion in 2001.

Although it is impossible to define what the optimal capacity is for the sector, intuitively the airline industry seems to suffer from capacity problems. In terms of excess competition, there is no history or technical analysis upon which to base a judgement. And the rate at which a new technology is adopted is difficult to forecast. But importantly, the physical technology of telecommunications is durable and the tendency is to write-down capital debt until a viable return is earned – the classic suburban rail problem. This may indicate an empty core but is somewhat different from airlines where capital is mobile, technology is relatively stable, and there is still significant investment taking place.

Many industries function for considerable lengths of time despite not recovering their full costs of capital. In some cases this may be possible because capital debts are written down through institutional means. The Chapter 11 bankruptcy laws in the US effectively allow a company to be restructured, and its capital written down, without the entity being broken up (see Annex II). Several

US carriers have made use of this mechanism (some on more than one occasion). The recent cases of America West, US Airways, and United Airlines are examples of this. From a wider perspective, competing firms often see this as unfair competition because the costs of their rival have been reduced through legal means rather than through strict efficiency improvements.

In other cases industries can attract capital not because the industry as a whole is seen as viable but rather on the speculation that some companies within it are. This effort to 'spot the winners' can provide a continual flow of investment even if the overall probability of making a profit is less than unity. This 'casino' effect can be heightened when an industry, or elements in it, have a particular appeal to investors for non-commercial reasons. Some sectors simply capture investors' imaginations for periods of time.

The most common case of this is perhaps sports team companies, but air transport also has a cachet about it that may attract capital even if it does not earn a reasonable financial return. Most industries are not independent entities but are part of larger value chains. In some cases those elements in the chain that are more than recovering their costs of capital may invest in other elements that are not recovering all costs. It is in their interest to do so if they can neither fulfill their role of the loss maker more efficiently themselves or if there are institutional constraints against them being direct providers. Airframe manufactures often support aircraft purchases with favorable loans to ensure adequate sales.

Industries also employ a wide variety of techniques to recover their full cost, or at least more complete recovery. Which is adopted depends upon a combination of the technical characteristics of the industry concerned and the institutional environment in which they operate. The main methods of cost recovery are examined below together with some practical illustrations of how they have worked in practice. There are always caveats attached to looking across industries and some of the main ones are outlined in Table 5.5.

Subsidies

Subsidies have long been used to recover the costs of capital. The argument is that once an investment has been made it is most economically efficient to maximize its use subject to the willingness of users to pay the incremental costs of their actions. The current trend is to unbundle attributes of a service and to attempt to isolate those where the fixed costs are concentrated. These costs can then be subsidized and the other attributes sold in the market at competitive prices. This has been a widely used policy for railways, with subsidized (and generally publicly owned) track being separated from commercially driven operations (Brooks and Button, 1995).

The UK's rail policy initially involved unbundling with track being placed under a single private authority while services were tendered for to keep operating subsidies to a minimum. The result has been a problem for the track authority (where most capital is tied up) to recover costs under a strict regulatory structure including capital subsidies. The operating companies have found it difficult to recover their costs even with local market power because of the unreliability of the track network.

A problem with any system of direct subsidies is that the incentive structure, unless the subsidies are allocated with care, makes it is difficult to ensure X-efficiency in production if attained. If the recipient knows that losses are to be financed from external sources there is no incentive to resist pay rises for labor or to economize on capital outlays. Further, there is much less of an incentive to provide the goods and products that customers seek. A variety of mechanisms have been developed to contain these potential intervention failures that can accompany direct subsidies.

Table 5.5 Caveats when comparing with other sectors

- *Deregulation has not covered a business cycle.* Airlines were at the cutting edge of the market liberalization movement; other sectors have only recently been subjected to competitive pressures.
- *Rapid technical change.* The airline sector is technologically progressive but far less so in recent years than sectors such as telecommunications.
- *Unbundling has taken place.* The privatization and liberalization of many sectors has involved separation of component parts of a sector (e.g., local from long distance telephone services). Air transport was largely unbundled before regulatory changes took place.
- *Regulations changed, not removed.* European airlines operate in essentially a free market, other industries have rather had the nature of regulations changed (e.g., from rate-of-return regulation to price capping).
- *No international/domestic mix.* Many industries serve only domestic markets, many European airlines serve domestic, intra-European and global markets.
- *Public ownership has only gradually been removed from airlines.* In many other industries the speed of privatization and its form have been different.
- *The output is durable.* Although not unique, air transport has the feature that its output is perishable.
- *Subject of speculation.* Other sectors for a variety of reasons have often been the subject of pure speculation.
- *Specific competition policy.* Different sectors enjoy a variety of exemptions from generic competition laws.

Institutionally, EU authorities frown on state aid for these types of reason and have been active in removing it from the airline sector. More recently, there has been aid given to airlines to attract them to fly to specific airports – Ryanair for example has according to EU calculus been enjoying subsidized landing fees at Charleroi airport in Belgium. While these types of practice are not uncommon in the US (e.g., Tallahassee and Wichita have been subsidizing the low cost carrier AirTran to service their communities for some years), the problems in the EU stem from the involvement of public monies. The fear is that while private industry's support for air services may reflect a genuine internalization of an external benefit and will be limited by market forces, there is no such mechanism to ensure that public funds will not be captured.

Natural market power

Natural monopolies can recover their full costs by dint of their market power. In technical terms they have the ability to either set their prices or their outputs to ensure that revenues exceed costs. The standard analysis assumes that they establish an output where marginal costs are equated with marginal revenues and then sell this output at a price that clears the market. But revenues can be even higher in circumstances where price discrimination is possible. At the extreme case where each potential customer is isolated the average revenue can be collected – hence the airlines' attraction to yield management.

Natural monopolies are, however, by their nature uncommon and are largely confined to the extractive industries. This is not only because there are few cases where an incumbent supplying a particular goods or service has no competition for that good or service, but also because good and services can themselves be substitutes – e.g., air travel can be substituted in some cases by rail services.

Some network providers (e.g., in telecommunications and energy) enjoy quasi-natural market power because the physical network is too expensive for duplication. In the US telecommunications market, for example, this has provided the traditional local suppliers (the Baby Bells) that retain control over the 'last-mile' of the system with the ability to extract reasonable returns. Long distance carriage, which is a highly competitive market, has found many suppliers with problems in full cost recovery.

In practice, there is always the tendency for governments to intervene when there is the potential for a natural monopoly to arise. This may be through regulations or public ownership. Whatever the case an inevitable trade-off emerges between containing the possible inefficiency and exploitation that can accompany natural monopoly power, and the potential intervention costs that can accompany government involvement in markets.

Institutional market power

Institutional monopoly comes about through such things as licensing and the tendering out of services. In some cases the structure is explicit, such as the licensing of a set number of taxicabs in a city or the tendering out of a bus service, but in other cases it may be in the form of self-regulation. The latter has been common amongst the professions such as medicine and law. Effectively, the government gives over the responsibility of limiting the degree of market entry to the industry itself. This in turn regulates who supplies the service and at what price. Adjustments can be made to ensure full costs are recovered. One of the arguments for doing this is that considerable skill and information is needed to ensure that those in the market are qualified, and it is more efficient for professionals to use their resource base to determine this.

The application of this type of self-regulation to airlines to assist in the sustainability of the market is unlikely to be acceptable to policy-makers as a way of ensuring cost recovery. It is used, however, in some air transport markets for things such as customer's rights, and IATA has fulfilled this self-regulatory role in a number of ways over the years.

Market power may also arise when suppliers merge or a dominant player exists. The UK 1986 Transport Act privatized and deregulated urban bus services outside of London. The former municipal fleets were sold off in small units on the mistaken premise there were no economies of scale. As with scheduled air services, however, the scheduled bus fleets found it difficult to recover costs despite adopting a wide range of tactics. As a result of subsequent mergers, one carrier – Stagecoach – now dominates the UK urban bus market. This has brought much greater stability to the industry and more reliable services to the consumer.

To prevent an institutional monopoly exerting excessive market power it is inevitably regulated. Traditional rate of return regulation that was used for many years in the public utilities allowed prices to recover all costs together with a reasonable rate of return. The limitation of this approach was that it was frequently captured by the industry under regulation. Essentially the industry had control over cost information and there was no incentive to keep costs down when any increase could be passed on to customers.

In some cases, workers or suppliers of inputs further exploited this because they knew there was no incentive for management to fight cost increases. More recent regulatory policies (e.g., as applied to BAA) have set price caps but in doing so they have allowed some costs to be passed on in price increases. In the UK this type of regulation is widespread and applied to telecommunication, energy utilities and to water.

Internal coalitions

Since the inability to recover costs is frequently due to the excesses of competition, the suppliers in the market may act to reduce this by forming vertical coalitions. At the extreme this may involve efforts to merge or take-over so as to, in effect, develop a monopoly as discussed above. Much more common are various forms of alliance. These provide a basis upon which the nature of competition can be adjusted to allow higher returns to be earned. The creation of shipping conferences in 1870 was, for example, the first major effort to create internal coalitions within international transport.

Alliances are now common in the airline industry and even extend to some limited alliances between airports. They appear beneficial to the airlines (see again Chapter 3). The reforms of freight rail transport in the US have resulted in a number of major alliances including those across the US/Canadian and US/Mexican borders. The outcome of these rail alliances has been capacity rationalization (10,000 miles of track between 1987 and 1999) and a significant growth in productivity and operating income in the 1990s (Transportation Research Board, 2002). A problem, however, is raising funds for investment in new services whilst rationalization is taking place.

Alliances often require legal approval because they have the potential of violating many tenets of competition and hence may violate generic anti-trust regulations. In most cases, however, companies have to raise revenues directly from their sales to recover their full costs. This is done in practice in a variety of ways but one thing is common to many of them. The firm needs a degree of market power to generate sufficient revenue. This type of issue is not always part of anti-trust thinking where ideas of potential excessive market power form the core of arguments rather than the desirability of some degree of market power to allow long term commercial viability.

Long term contracts between supplier and customer

By negotiating a long-term cost recovery contract with a major customer, at the time capacity is introduced, a supplier can ensure that there is a guaranteed revenue flow that will cover most of the capital outlay. This is often standard practice in such traditional industries as steel production where long term agreements are reached with customers. The key features here are that the investments tend to be very lumpy and relatively long lived, and that there are not many potential customers.

Such arrangements are not uncommon in the air transport field. Perhaps the most discussed is the Civil Reserve Air Force (CRAF) of the US whereby the federal government gives exclusive contracts to US carriers for the carriage of their personnel on condition that aircraft and crew are available at times of national emergency. Similar arrangements have helped in many European countries. Another example is long-term contracts between post offices and airlines to carry mail. Such contracts are also found outside of the airline sector, for example between railroads and power stations for the regular carriage of coal. Branch lines are constructed on the basis that coal will be carried at a pre-determined rate for a contracted number of years.

Advanced revenue with subsequent capital adjustments

Most business activities involve making an investment with fairly predictable costs and then seek to repay this from much less certain revenue flows. An alternative is to try and secure a more certain revenue flow and then to adjust capital outlays so that a viable return may be earned. This is in effect what many charter airlines do. They sell capacity in advance to tour operators and are fairly well informed many months ahead of when their capacity has to be delivered. Freight railway companies in Canada and the US, when carrying seasonal products such as wheat, pursue similar practices.

Scheduled airlines cannot follow this pattern because they guarantee a service ahead of time and then effectively become common carriers of the traffic willing to pay for flights. In some US cities groups of businessmen have tried to attract carriers with guarantees of adequate patronage for an initial period. These are known as 'travel banks'. In Wichita some 400 businesses raised $7.2 million to attract carriers. Air Tran started operations in May 2002 with services to Atlanta and Chicago's Midway airport. It will get up to $3.0 million to cover losses in its first year and $1.5 million in the second. Similarly, Pensacola raised $2.1 million from 319 businesses to attract Air Tran while companies and individuals in Stockton bought $800,000 of prepaid tickets to attract American West.

The full-service airlines engage in this type of long-term revenue management through frequent flier programs. These are, amongst other things; intended to reduce fluctuations in the revenue flow and thus inject greater certainty into longer term planning.

Two-part tariffs

Separating out capital from operating or marginal costs has been a standard way for many utilities to recover their full costs. In the case of telecommunications (be it hard wire or cellular) consumers normally pay an access fee (roughly equal to the capital cost) and a usage fee. In some cases, if the marginal costs are extremely low, there is simply an access fee that subsumes the estimated costs of individual calls. This is a standard practice for local telephone services. The two-part tariff is also to be found in a variety of private sector activities – for example the golf club membership and the green fees.

The successful use of two-part tariffs depends very much on both the relative importance of the capital element in costs and the frequency with which individual consumers use the good or service. It is generally far more successful when there is regular and heavy use. This is because the supplier can employ information to adjust pricing and investment plans and consumers do not feel a heavy burden when it comes to paying the periodic 'membership' fee. It is also most efficient when the service being offered is fairly homogeneous. This makes allocating the fixed fee easier. Its use in air transport is relatively limited because most people do not travel often enough to make it an attractive option and the nature of trip making means that it is difficult to allocate a reasonable approximation of capital costs to each consumer.

Vertical integration

If one link in the value chain fails to recover its full long run costs but the chain in its entirety is viable, then one option is for the loss-making element to vertically integrate with profitable links. Historically this was done in many transport industries when feeder services were vertically integrated with the mainline services – e.g., often the railway companies operated bus and trucking services to feed traffic to their mainline services. It is common practice for many large companies to self-insure to avoid the costs of outsourcing.

Vertical integration has taken a variety of forms in the air transport industry and is still practiced today. In the early days of aviation airlines were often vertically integrated with aircraft manufactures (Boeing and United Airlines being an example).

More recently, many airlines outside of Europe have made heavy investments in airport infrastructure. In terms of ticket distribution and information systems, airlines were involved in the development and use of computer reservations systems. In the US there was direct ownership of the systems until divestiture was thought to be a better commercial proposition. The development of Orbitz in the US and the somewhat less all- embracing Opodo in Europe reflects a partial return to this form of vertical integration. The difficulty with vertical integration is that it imposes additional managerial strains to the system. It essentially moves airlines away from their core business and reduces the efficiency with which they can determine their profit centers.

A case where vertical integration has been common has been the railways – although this is changing in Europe. In the US the Staggers Act of 1980 removed most economic regulations over the freight railways. The rail companies had control over track and operations. Competition pushed down revenues but reductions in capacity (which often involved selling non-core lines to smaller railroads), reduction in the numbers of workers, more productive labor practices, and containment over wage increases meant costs fell even more. In 1986 the railroad operating expenses were 93% of costs but by 1995 they were 86% of costs. Basically, the railroads had control over a major input cost – their infrastructure.

Discriminate pricing

Discriminate pricing is widely practiced in transport and other network industries. The underlying idea was summarized by the UK Office of Fair Trading (1999), 'In general undertakings will need to set prices above their incremental costs so that common costs, for example, can be recovered. Price discrimination between different customer groups can be a means of achieving this; it can increase output and lead to customers who might otherwise be priced out of the market being served. In particular, in industries with high fixed or common costs or low marginal costs, it may be more efficient to set higher prices to consumers with a higher willingness to pay.' Although as a recent Special Report of the US Transportation Research Board (1999) expressed in the particular case of airlines (although it has general applicability) 'In the long term, sustaining such a pricing scheme usually requires government regulation or market power to bar entry.'

Discriminate pricing was developed and refined in the US domestic air transport market in the form of yield management. This is basically dynamic discriminate pricing in that the availability of service at any fare is changed as a plane fills. Similar techniques are used in the scheduled maritime container industry and peak and off-peak fare differentials are common for local public transport systems. Outside of transport, deregulated long telecommunications providers give quantity discounts to both large and small customers; charge business and individuals different rates; and offer calling plans that offer discounted rates based on individual characteristics and usage patterns.

The idea of yield management is to extract as much revenue from customers as possible by levying prices that reflect the willingness of customers to pay. Consequently, customers that are less sensitive to price pay more and contribute to the capital cost of the service, whilst those who are less willing to pay are charged lower prices that at least cover their marginal costs. It has been long used in a particular form to recover the capital costs of various public utilities such as passenger rail services, urban bus services, and energy. Here the aim is not to price differentiate between users so as to maximize profit but rather to generate revenue so that an acceptable return is earned after all costs (including the cost of capital) have been covered. This approach, known as Ramsey pricing, constrains the level of prices levied at the upper end.

Relevance for the airlines

While there are many ways in which full cost recovery is possible, the perishable nature of airline services and the lack of market power enjoyed by carriers often makes many of them inappropriate for airlines. Indeed, in many cases they have not proved successful in the industries where they have been adopted. What does seem to emerge is that there are ways of mitigating the instability problem but no single solution. In the short term there can be a greater focus on those services where returns are highest and, indeed, recent years has seen considerable restructuring of the networks of the several full-service carriers to this end – e.g., see Button et al's (2004) discussion of developments at TAP. Correction of the value chain would both clarify the overall optimal capacity of the industry, and remove some of the distortions to competition.

Changes in attitudes to competition policy giving airlines the ability to more easily form alliances and mergers would result in greater control over market conditions and investment, and permit a more structured policy towards the no-frills carriers without a return to a heavily regulated environment. In some cases this needs little more than a de facto reinterpretation of existing laws.

The airlines themselves can also be more proactive, and in some cases there is evidence that this is taking place. A limited amount has also been done in terms of vertical integration – e.g., the development of Opodo and the exercising of countervailing power by some of the airlines in airport negotiations. And airlines already are active in such measures as yield management and customer retention programs to smooth out fluctuations in demand, although the effectiveness of this seems to be waning. Mergers to create more viable networks offer another small step at self-help. It seems unlikely, however, that such measures will by themselves resolve all of the issues.

Chapter 6

Conclusions

This book has looked at the need for, and the importance, of developing an efficient commercial air transport system for Europe by addressing two fundamental questions:

- Is scheduled air transport important for Europe?
- Is the current structure of scheduled airlines in Europe sustainable?

The book has not offered a detailed study of the short-term financial problems being encountered by many European air carriers, but has looked mainly at the underlying structure of the airline market as it moves into the 21st Century. Nor is it a study aimed at producing a plethora of new statistics. Rather it seeks to highlight some of the challenges that exist, and seem likely to be continuing problems for European airlines. The airline world is in a continual state of change, and inevitably policy changes will be required by both government and by those in the industry.

The objective here, however, has not been to offer policy solutions but rather provide a deeper understanding of the dynamics of the industry, as a basis upon which long-term policy may be developed. The style of presentation is consequently more academic than is often found in business reports. It extends beyond just giving a snap shot of current data to look at the wider perspective. It is also about the airline industry and not about the individual carriers involved – an industrial perspective often differs from that of individual firms.

Air transport is important

Air transport allows for personal mobility and for the speedy movement of goods. It is a large global industry (carrying over 1,600 million passengers a year and doing more than 130 billion revenue ton kilometers of cargo). Importantly it is a very rapidly growing industry and both passenger and cargo traffic is forecast to expand faster than the global economy into the foreseeable future. This reflects the importance society places on it.

Within Europe commercial air transport is a vital form of medium distance transport and ties Europe into an intercontinental market place. Its function is that of a lubricant in the modern world of trade and commerce where speed, economy, and reliability of transport are crucial for the continuation of a region's comparative advantage.

Major airports at the hub of scheduled air transport networks provide a catalyst for national, regional, and local development. Air transport can be particularly important in stimulating the growth of modern, high technology industry in an area both by facilitating interaction of professionals and management, and by offering the types of service required for just-in-time production techniques. It also provides the main transport mode for much of the European tourist industry that is the backbone to many European economies.

As the European Union expands geographically as a result of the Nice and Copenhagen agreements, and deepens with the move towards agreeing on a constitution it is inevitable that air transport networks will be central to the integration of new members, and to the ability of the enlarged area to reap the full external economic potential of scale and internal economies of comparative advantages.

A new market structure of European air transport is emerging

The global move towards freeing air transport markets from economic regulation has produced short-term benefits for many consumers of airline services. These have included innovative services, more flexible and generally lower prices, and more choice of suppliers. Internationally, the liberalization of the European market came relatively late and was phased in which initially was seen to be less disruptive than the sudden changes often imposed elsewhere.

The process is still incomplete in that the external international air transport relations of the EU are not negotiated in a coordinated manner but involve numerous bilateral treaties negotiated at the national level. Internally, there are still issues of coordination and standardization of air traffic control systems and of such things as airport pricing and investment policies. But what is clear is that the European airlines, despite the phased nature of liberalization, are at the beginning of the 21st century experiencing similar long-run commercial challenges as their US and other counterparts.

The full service European airlines have been adjusting their business plans in response to these institutional changes and as a reaction to new technologies, managerial concepts and emerging consumer trends. The full service carriers, now often privatized, have entered into more and larger strategic alliances, often involving participation in a major global network. They have developed hub-and-spoke systems that provide basic air services to communities on thin routes as well as major cities.

Some of the long-standing European carriers such as Sabena and Swissair have found it difficult to integrate into this structure and have gone bankrupt. There are new types of airline emerging – the no-frills scheduled carriers. These have, by offering low fares and basic services, opened up new markets, but also offer competition on some routes with established scheduled and charter carriers. Their point-to-point services enable low operating costs but do not allow interconnectivity of the airline networks.

The institutional environment has changed

The European airlines have had to adjust to major institutional change at the same time as encountering new market conditions and handling the more traditional ones of trade cycles and unexpected external shocks. The decade of liberalization of the internal air transport market of the European Economic Area, coupled with the less systematic changes that have taken place in markets outside, have required major structural changes. Airlines have had to move to a commercially driven environment as a succession of Packages has removed many of the elements of economic regulation from the Single European Market.

The industry has come more directly under the competition policy of the European Union. Outside of the European Economic Area, the largely US driven Open Skies regime has fostered more open entry into many routes whilst at the same time providing an institutional structure in which strategic alliances can emerge. Many parts of the world, however, still operate under restrictive bilateral regimes.

Changes are having diverse implications

The observable impact in the short-term has been a demonstrable increase in the economic efficiency of the full service carriers in Europe that has been passed on in lower fares and more convenient services to travelers. The significant secular increase seen in the amount of air travel undertaken, which has consistently exceeded income growth, points to the revealed preferences of the citizens of Europe to fly.

As with the US and many other scheduled markets, traditional network airlines are currently encountering short-term financial problems stemming from the immediate aftermath of events in New York and Washington in September 2001. This is not surprising given the immediate effects on air travel demand, subsequent impacts on economic growth and added costs of airline security to the system. They are also taking longer to recover than was initially foreseen and the subsequent outbreak of SARS and added security concerns after the military action in Iraq have added to airlines' problems.

While the US airlines are experiencing the most serious financial problems, the European carriers are at a longer term disadvantage because of the significantly smaller compensation packages they received and because European bankruptcy laws (or at least their implementation) make it more difficult for carriers to restructure their finances.

These immediate financial difficulties, however, simply amplify longer-term structural problems in the scheduled airline industry in Europe and elsewhere. Taking a long-run view that covers the full business cycle or, in the case of the US, several business cycles there is clear evidence that full service carriers offering services in liberalized markets are not generating sufficient revenue to cover their full capital costs. Their net profit margins are by normal commercial standards very low.

There are distortions in the value chain

A feature of the airline market contributing to its problems is its atypical structure within the longer supply chain that leads to air service provision. Whilst airlines earn about a 4% to 6% return on capital invested, compared to a return of about 9.5% considered necessary to cover full capital costs in the early 2000s, many suppliers of materials and services used as inputs by the scheduled airlines enjoy considerably higher returns. Institutionally, the difficulty is that while the airline market has been liberalized, and competition policy has been exercised in a traditional way to contain any market power, there are serious market failures elsewhere in the value chain.

Many of the supplying industries, such as aircraft manufacturers, global distribution systems, and catering operate in markets that are duopolies or have very few companies. The involvement of large financial service companies at some stages in the chain also provides resources to cover the worst period of downturns in the business cycle. In other cases, such as air traffic control and airports, there are natural monopolies that have little incentive to provide efficient service or keep prices charged to airline customers down. The structure of airline labor markets gives strong bargaining powers to small numbers of workers allowing them to extract high pay.

Airlines are not covering their full costs

These factors make it difficult for the full service airlines to cover their full costs, and the problem of continuing legal and political contracts inherited from the days of public ownership and regulation does not permit flexibility during downturns in the business cycle. In themselves, however, they would seem to explain only part of the problem. While high input costs, legacy effects and the on-going need to adjust to the new commercial environment inevitably lead to higher costs, this should not in itself mean that those, albeit higher costs cannot be recovered. It rather means that the overall size of the industry is likely to be sub-optimal which is detrimental to the long-term welfare of consumers. Indeed, the full service carriers in the US domestic passenger market, although having a number of important distinguishing features, repeatedly experience the same type of problem. And that market has been deregulated for nearly a quarter of a century.

A number of explanations have traditionally been put forward to explain this type of situation. These often include the existence of substantial amounts of stranded capital costs that the supplier cannot recover. This may be because of declining demand or the existence of inappropriate incentive structure confronting management – problems usually associated with public works. None of these features would seem applicable in any significant degree to the European airline market. Demand is certainly not falling in the long term; indeed it has been and would seem in the future, if the forecasts are even remotely accurate, very robust. The liberalization of the airline market, the significant amount of privatization that has taken place and the vigor of the European competition authorities give little scope for managerial slack.

Similar issues exist in other industries

Other expanding industries have historically appeared prone to capacity problems. The scheduled shipping industry initially had this type of problem but rapidly managed to contain it by obtaining the legal right for liners to collude on matters such as schedules and capacity – the conference system that has now developed into the consortia structure with the advent of containers.

More recently the deregulation of energy and telecommunications markets about the globe have seen cases of rapid expansion and over supply in the short term. These, however, have been in sectors experiencing considerable technical progress and involving the experimentation of new forms of finance and public policy. Unlike airlines, where the US, and other markets offer insights over entire business cycles, many of the difficulties in these sectors may be attributable to severe bubble effects immediately following liberalization. But again there are lessons to be learned.

Market pressures push airfares below full costs

A major problem with scheduled airlines is that in competitive environments there are natural market forces tending to push fares down to marginal costs. The airlines have, although perhaps not consciously on an individual basis, sought to contain these pressures by preserving their markets through loyalty systems, controls over information flows, and alliances with other carriers. Through yield management they have sought to extract what consumers are willing to pay for a service. These strategies are encountering problems as new information systems, the power of other actors in the value chain, and the controls of regulatory authorities have limited the abilities of carriers to confront the challenges of highly competitive conditions when committed to provide a fixed schedule of output.

No-frills carriers have an important place in the market

No-frills carriers may be seen as adding to instability problems, especially when there are any cross-subsidies in the system favoring them (e.g., as between airports). They must be seen, though, as meeting the needs of many customers and it seems likely that they will expand in importance, albeit at a slower rate than in the past. Some no-frills carriers have developed business models that fit with significant niche markets. The issue for the full service carriers and for European policy makers is to develop a structure that combines the incentives inherent in competition without pushing prices down to the point where it is impossible for suppliers to recover full costs.

The current situation is unsustainable

The existing system of providing air transport in Europe is sub-optimal in that monopoly elements elsewhere in the value chain result in a distortions to the scale of the airline market and the impediments to full competition are such as to prevent full cost recovery. Such a situation could limp on for some time but in the long run is unlikely to prove attractive to air transport service users, the financial markets, or governments.

Annex I

European Union Air Transport Policy

The early years

While there has been a Common Transport Policy since the signing of the Rome Treaty in 1957, aviation was initially excluded from EU policies.[1] Countries regulated their own domestic aviation and a bilateral system of agreements, evolved from the Chicago Convention of 1944, governed international air transport within the Union, as well as outside it. Policies were concerned with the regulation of scheduled fares, service provision and market entry. Coupled with this was the growth of a very large European charter market that largely met the demands of rapidly expanding north-south tourist traffic. This market was less rigorously regulated and served by low cost operators.

The EU has never had a single regulatory body with responsibility for international air transport like the former Civil Aeronautics Board in the US. The bilateral air service agreements that emerged after the Chicago meeting were piece–meal arrangements, although common motivations often led to standard features. They were generally restrictive and often allowed only one airline from each country to operate a route. Over 90% of bilateral agreements involved controlled capacity with obligatory 50:50 revenue pooling. Some 900 of the agreements excluded 5th-freedom rights. The two countries involved agreed on the airfare, and competitive pricing was excluded. Many of the airlines were state-owned flag carriers and received substantial state subsidies.

Within the EU, overlapping philosophies to economic regulation extended into air transport. The patchwork of controls over market entry, fares, and conditions of operation that existed in 1957 grew with time reflecting these differences. Countries such as France, Spain and Greece, where domestic aviation is relatively important, have a tradition of heavily regulating entry and fares, and this extended to their views of international aviation policy. There was also a pervasive philosophy that air transport serves public needs and that to both ensure adequate provision and to avoid the economic distortions of monopoly power state ownership best served the public interest.

This degree of regulation and state involvement was thus frequently justified by governments in terms of serving the public interest by ensuring market stability, maintaining safety standards, protecting the public from monopoly exploitation

---

[1]   The term European Union is used here to reflect current terminology. In fact, there have been several changes in title and in broader institutional structures of what is now the EU. The term is adopted here for simplicity.

and providing a comprehensive network of services. Regulations also served as instruments for the protection of other aspects of the national interests by maintaining flag carriers to meet wider economic and military criteria. Exporting aviation services can also represent an important element of 'invisible' earnings from foreign trade. In addition, there was the question of status and market presence. In some countries (e.g., Greece), aviation was provided through statute by national, state-owned airlines. Such direct controls not only influence air transport services but can also be deployed to regulate the purchase of aviation equipment which can also form a major item in foreign trade accounts.

Such a regime of ad hoc, state-based regulations is unlikely to generate an efficient air transport system. While some countries may have benefited because of their bargaining position or through historical accident, overall it protected inefficient operations and distorted the overall pattern of services. It acted as a restraint on trade and had associated with it the same economic implications as tariffs. The problem was that countries with well-entrenched systems of market controls, even if appreciative of the probable adverse implications of this for the overall welfare of the Union, still effectively cushioned their airlines from competition and found it difficult suddenly to compete in a more market-oriented environment. Change came slowly and came in several ways.

The incremental changes

Reform of the European internal market materialized as a series of steps. Whilst the later stages saw explicit EU involvement and a series of conscious policy initiatives, the earlier reforms were very much more of an ad hoc nature. They largely relied upon immediate self interest of the parties involved, or a degree of bilateral reciprocity of interests.

Domestic reforms

There were initially reforms of domestic policy within some EU member states. Some, such as in the UK, were de facto changes in the interpretation and implementation of existing laws, and as such did not entirely free up the market. They saw the national regulatory agency being more liberal in the allocation of licenses and acceptance of fare flexibility (UK Civil Aviation Authority, 1988). Some countries – e.g., France, Spain, Italy, and Germany – were less inclined towards unilateral domestic liberalization.

Reforms were gradually accompanied by greater private sector involvement in the sector. In some instances (e.g., British Airways), there was complete privatization of former state companies at an early stage. More common (in Germany and the Netherlands) was a gradual selling off of stock. Airports and other fixed infrastructure, outside the UK where main airports were privatized as the British Airports Authority in 1987, have tended to remain in the public sector.

Reforming bilateral air service agreements

Since the mid-1980s, there has been a move to liberalize bilateral agreements between some members. In 1984 the UK and the Netherlands concluded an agreement that significantly relaxed the rules covering traffic between the two. The main features were that any airline (based in either country) could fly between the two signatories and that tariff freedom was established: there was no compulsory consultation with other airlines and fares were to be set by the airline of the country from which a flight originated. The capacity offered was left to the airlines. In 1985 the countries introduced a double disapproval system. Previously fares had been subject to approval by the country of origin.

But from 1985, airlines were free to modify fares unless both countries disapproved. Subsequently the Anglo-German (1984), Anglo-Belgium (1985), Anglo-Luxembourg (1985), Anglo-Italian (1985) and Anglo-Irish (1988) agreements all embodied varying degrees of liberalization. Whilst these developments were a step forward, what was still lacking was a Union-wide act of liberalization. This was something that could only be accomplished by the Council of Ministers.

The changes did initiate liberalization into fragments of the EU international market. It encouraged new entrants and brought response from incumbents. Evidence from Irish–UK routes point to lower fare levels and benefits estimated in 1989 at £24.9 million for the 994,000 existing passengers and £16.2 million for 1.3 million additional passengers generated post-liberalization (Barrett, 1990). The impact of the Anglo-Dutch liberalization was more muted which can in part be explained in terms of its timing in the trade cycle and the fact that the existing bilateral was not excessively restrictive.

'Open Skies' agreements

Not all liberalizing measures were exclusively within the Union, such as the Anglo-Swiss bilateral free access capacity provisions with limited tariff constraints. Of particular importance were agreements involving the US whose 'Open Skies' policy, since 1979, has attempted to develop liberal bilateral air service agreements with individual EU states. The bilateral agreements between the US and individual EU countries had traditionally varied by country but in general were relatively restrictive on capacity and entry points as well as 5th freedom rights. There was a gradual change in the 1990s but, to date, the only long-standing liberal agreement is with the Netherlands.

In 1994 the US initiated liberal bilateral agreements with a number of European countries, although none had a major international airline. In 1996, an interim arrangement was reached with Germany amounting to an Open Skies agreement. The impact of both types of development and the stance the US has taken is to help bring European aviation closer to the global marketplace. The liberalization of the North Atlantic has not been entirely painless in the short term. Between 1984 and 1990 the six countries with the most liberal bilateral air service agreements with the US (Belgium, Denmark, France, Germany, Spain and the Netherlands) lost market share while those with more restrictive agreements (Greece, Denmark, Italy, Portugal and the UK) gained market share. Over the same period, overall the US carriers took a larger proportion of the fast-growing non-US citizens traffic on the Atlantic.

A longer standing issue, in the context of an EU air transport policy, is whether it is still appropriate for individual member states to engage in such bilateral arrangements or if a common EU negotiating position should be taken (Association of European Airlines, 1999). In 1996, EU states gave over some soft-negotiating rights to the Commission but major players in the transatlantic market, and especially the UK, have resisted relinquishing their individual negotiating rights.

The early developments of EU policy

The EU's air transport policy must be put into context. What have become the foundations of the regulation of scheduled air passenger services were laid out shortly before the end of World War Two at the Chicago Convention. Inter-state services in Europe were no exception. The Convention gave rise to a United Nations agency – the ICAO – that was largely concerned with technical standards, the collection of statistical data, etc., rather than detailed economic regulation. Governments claiming absolute sovereignty over the airspace above their territories dominated discussion at the Convention. The underlying hope was that those signing would grant to all other signatories freedom of access to this airspace and to airports beneath them. Failure to achieve this resulted in a system of bilateral air services agreements, many of which were highly restrictive and protectionist.

The bilateral air service agreement regime

Bilateral air service agreements involving EU members were inevitably not uniform, but there were typical features.

- Access to the market was not free but was severely restricted. Often only one airline from each country was allowed to fly on a route. This was referred to as single designation. As late as 1987, out of 988 routes within the Union air services network only 48 had multiple designation, i.e., more than one airline from either side.
- Fifth freedom competition was the exception – of the 988 routes only 88 allowed 5th freedom rights. The capacity offered by a bilateral partner was restricted – generally each state enjoyed 50% of the traffic.
- Pools in which the airlines shared the revenue in proportion to the capacity employed often accompanied the division of the market. Thus even if one airline obtained 54% of the revenue and the other thus enjoyed only 46% they would nevertheless split the proceeds equally.

- The regulatory bodies of bilateral partners approved fares and there was no competition on price.
- The airlines designated by a country had to be substantially owned and controlled by it, or by its nationals.
- Some airlines, many wholly or partially government owned, enjoyed competition distorting state aids.

Air transport and the Rome Treaty

The 1957 Rome Treaty establishing the European Economic Community, effective from January 1958, originally involved West Germany, France, Italy, the Netherlands, Belgium and Luxembourg. Subsequently it was enlarged – in 1973 the UK, Ireland and Denmark became members of what became the European Community. In 1981 Greece was admitted and in 1986 the Community was enlarged to include Spain and Portugal. Sweden, Austria and Finland are now members.

The Treaty required free competition across frontiers calling for the removal of all protective tariffs, quota and non-tariff barriers to the free circulation of goods. The Treaty envisaged free competition across frontiers for services. Around the free trade area there was to be a common external tariff. The EEC envisaged free factor mobility.

Three features of this arrangement had potential for undermining the system of air transport regulation. Article 3 of the Treaty stated that the objective of the EEC and one of its key aims was the creation of conditions of undistorted competition.

This was in conflict with policies in several sectors, including aviation. The system of airline regulation was anti-competitive and fundamentally at variance with this competitive ethos.

Secondly, the requirement to abolish non-tariff barriers applied not merely to goods but also to services. Antitrust-type restrictions constitute such a barrier and the European Communities Commission could ban cartel arrangements. The price-fixing and revenue pooling of the airline regulatory system appeared to be extremely vulnerable to attack.

Thirdly, a key feature of the bilateral agreements was the restriction on access to routes that were at variance with the idea of freedom to supply services. But, whilst the Rome Treaty appeared to threaten the system, the Community made only limited progress in removing the restrictions on the free movement of goods, services and factors.

At the outset, the balance of forces favored the existing regulatory system. While the Commission was charged with the implementation of a treaty that embraced a pro-competition philosophy, the competitive rigor of the Commission's policy was conditioned by the prevailing orthodoxy. Moreover the Commission was faced with some uncertainty over the position of air transport under the Treaty. This provided for a separate regime for transport that could be somewhat different from the rest of the Treaty.

However, whilst the Rome Treaty was sparing when it came to spelling out the detailed nature of a Common Transport Policy (CTP), the provisions that were revealed were liberalizing. Moreover when the Commission came to put policy flesh on the Treaty it chose to adopt a competitive stance. However, the CTP only applied to road, rail and inland waterways – air and maritime transport were not included. Technically, the Council of Ministers was left with the power to decide what provisions should be made for air and sea transport.

Other players were also favorable to the status quo or were relatively quiescent. Member states had created the regulatory system and the prevailing orthodoxy plus inertia did not dispose them to undo it. Scheduled flag carriers enjoyed the protection of the system and were unlikely to deviate from the position taken by their governments. Free and intensified competition carried with it the possibility of losses that would fall upon national exchequers. The non-scheduled airlines were carving out a role for themselves but they had not gained enough confidence to mount a direct assault on the preserve of the scheduled carriers. It was only as air travel grew that users became an increasingly significant pressure group.

Airline policies

In 1974 the first significant change occurred. In the French Merchant Seamen case the European Court of Justice declared that the general rules of the Treaty, and by implication those relating to competition, applied to sectors such as air transport. This appeared to open up air transport regulation to attack. However, this was not so. The Treaty allowed the Commission to prohibit activities such as collusive price fixing and abuses of dominant positions but even if applicable to air transport there was still the problem that the Commission lacked powers to directly implement these rules. The Commission also lacked powers to directly impose penalties.

The 'Memorandum'

In 1981 attempts were made to take the Commission to task for not applying competition rules to fare-fixing restrictions. The Commission replied that fare fixing was an autonomous act of government while the competition rules applied to enterprises. A further difficulty arose in connection with designation. Licensing often only allowed one airline from each bilateral partner. This appeared to conflict with the freedom to supply services. However, Article 90, which covered public enterprises and enterprises to which member states grant special privileges, said that the rules of the Treaty should not be applied if their application would obstruct these enterprises in the tasks assigned to them. Members could therefore declare that their national flag carriers were enterprises granted special privileges and that strict licensing principles were essential.

Ultimately, the changing climate of public and political opinion had an effect. The Commission took a bolder line in 1979 and issued *Civil Aviation Memorandum No. 1.*

- Increased possibilities of entry and innovation were desirable but freedom of access was a long-term prospect.
- There was a need for the introduction of various forms of cheap fare.
- There was a need to develop new cross-frontier services connecting regional centers within the Community.
- An implementing regulation applying Articles 85 and 86 directly to air services was essential – a proposal on these lines was made in 1981.
- Increased competition emphasized the need for a policy on state aids.
- Whilst the right of establishment applied directly to airlines, Council action was necessary since practical and political obstacles would otherwise still exist.

In 1984 came *Civil Aviation Memorandum No. 2* that moved on to more specific liberalization proposals.

- Fares should be subject to a zone of flexibility system. A reference fare level and a zone of reasonableness around it would be arrived at on the basis of official double approval – i.e., both sides would have to agree. Having thus determined the scope for flexibility, airfares within the zone of flexibility would be subject to country of origin approval or double disapproval.
- The dominance of national flag fliers within the bilateral system was to remain relatively undisturbed except for allowing other carriers to enter and take up any unused route operating rights.
- In the case of the 50/50 division of traffic between the national flag carriers, state A could not oppose a build up of traffic by state B until state A's share had fallen to 25%.
- Inter-airline agreements should be subjected to control. Capacity agreements would be permissible if airlines were free to withdraw. Revenue sharing pools might be exempted if their redistributing effect was minimal.

Empirical studies and legal cases

The European Civil Aviation Conference (1981) showed the restrictive nature of competition. It found that only on 2% of city-to-city routes was there more than one airline operating per state. On 93% of routes there were limitations on the number of flights that airlines could put on. Revenue pooling arrangements covered some 75% to 85% of total ton-kilometers flown. Comparisons between airfares in Europe and in the US were made by the UK House of Lords – certain fares were double the US fares for similar distances, although the differences for return fares were much less marked.

Other studies emerged supporting the view that European airfares were significantly higher. The Commission of the European Communities (1981) carried out investigations. Just as scholars in the US had been able to compare the free competition fares of California with the CAB regulated rates, so it was possible in Europe to compare bilateral regulated rates with those offered by the competitive charter sector. Barrett (1987) reviewing the European Commission's conclusions based on cascade studies draws attention to the Commission's observation that 'it would appear that only a relatively small proportion of the difference between scheduled and charter costs cannot be attributed to inherent differences between the two modes of operation'.

There were two final precipitating factors. In 1979, the European Parliament was for the first time directly elected. This enhanced its authority and it put forward a far-reaching draft Treaty of European Union. This led to an intergovernmental conference. Whilst the member states did not agree to go along with Parliament's scheme, they recognized the need to revive the idea of closer economic union. This led to the 1986 Single European Act a key feature of which was the economic commitment to complete the internal market by the end of 1992. Part of the program involved the need to address the continuing restrictions in air transport.

The final push came from the European Court of Justice and the Nouvelles Frontières case concerning the air fare-cutting activities of, amongst others, a French travel agent. Tickets had been sold below the officially approved price – a violation of the Code de l'Aviation Civile. The legal proceedings in France generated an appeal to the Court of Justice as to the conformity of the Code de l'Aviation Civile with the law of the Community.

The result of the judgment was to encourage the Commission in the view that its powers to attack fare-fixing activities were greater than the lack of an implementing regulation might suggest. The Court said that if the Commission, or an appropriate national authority, was to pronounce adversely on an airline restriction, then a national court would have to take account of that fact. A party with standing might therefore bring an action against such a restriction in a national court and the restrictive arrangement in total might fall.

The 'Three Packages'

The major reforms to EU policy came as 'Three Packages'. The details of these were essentially agreed separately and their time was not predetermined. They represented a gradual phasing in of reforms.

The 'First Package'

The outcome of the legal challenges, the demonstration effects of regulatory change in the US and the implicit move to a Single European Market, forced the Council of Ministers to act although there were differences between Members on the desirability of competition. The Commission decided to increase pressure and to instigate proceedings against certain airlines. The Council decided that the best way to regain control was to agree to introduce deregulation but of a kind, and at a pace, of its choosing. Hence the 1987 deregulation package – the 'First Package'.

The basic philosophy of these measures was that deregulation would take place in stages – evolution rather than revolution being the watchword and workable competition being the objective.

A regulation was adopted enabling the Commission to apply the antitrust articles directly to airline operations. Only inter-state operations were covered, the intra-state services and services to third countries were not affected by this measure. Certain technical agreements were also left untouched.

Whilst this enabled the Commission to attack agreements restricting competition (and abuses by firms in a market-dominating position) another regulation was introduced that allowed airlines to continue to collude in certain matters. This enabling regulation allowed the Commission to exempt *en bloc* three categories of agreements: concerning joint planning and coordination of capacity, revenue sharing, consultations on tariffs and aircraft parking slot allocation; relating to computer reservation systems; and about ground handling services. This block exemption power was of limited duration – it had to be revised by July 1990 and Commission regulations made under it expired by February 1991.

The Council also adopted a directive designed to provide airlines with greater pricing freedom. Whilst airlines could collude, the hope was that they would increasingly act individually. The degree to which competition emerged was recognized as depending on the degree to which airlines exhibited a competitive spirit in their approach to airfare applications made to national civil aviation authorities.

The directive declared that member states should approve fare applications provided they were reasonably related to the long-term fully allocated costs of the applicant air carrier, while taking account of other relevant factors. They had to consider the needs of consumers, the need for a satisfactory return on capital, the market situation, including the fares of other air carriers operating the route and the need to prevent dumping. State authorities were not allowed to keep price competition at bay by refusing to approve a fare application simply because it was lower than that offered by another carrier.

The new arrangements did not constitute free competition. Whilst conditions were laid down that reduced the national authorities' room for maneuver in rejecting airfares, they could still do so. If, however, there was disagreement on a fare the disagreeing party lost the right of veto because a right of arbitration was provided for and under this the disagreeing party could have its case overturned. The key fares in the approval procedures were economy fares.

The directive provided scope for discounts. Provided certain travel conditions are met, fares could be reduced below the benchmark by varying amounts. Discounting and deep discounting were automatic rights. There was an additional degree of flexibility for fares which, when the directive was introduced, already fell below the bottom of the deep discount zone. Fifth freedom operators could match these discounted fares.

The 1987 package also made a start on liberalizing access to the market. A decision was adopted that provided for a deviation from the traditional air services agreement's 50/50 traffic split between the two countries. Member states were required to allow competition to change the shares up to 55/45 in the period to 30 September 1989 and thereafter to allow it to change to 60/40. Fifth freedom traffic was not included in these ratios and was thus additional. There was also a provision in which serious financial damage to an air carrier could constitute grounds for the Commission to modify the 60/40 limit.

The decision additionally required members to accept multiple designation on a country pair basis by another member state. A member state was not obliged to accept the designation of more than one air carrier on a route by the other state (i.e., city pair basis) unless certain conditions were satisfied. These conditions become progressively less restrictive over time.

The decision also made certain 5th freedom rights automatic but these were hedged around with safeguards. Flights had to be extensions of a service from the state of registration or a preliminary of a service to the state of registration. An example of the latter would be that Aer Lingus might already have a right to fly between Birmingham and Dublin. It could now enter the Brussels to Birmingham route normally reserved for UK (and Belgian) bilateral operators, i.e. it could drop off in Birmingham passengers picked up in Brussels on a flight en route to Dublin. Additionally, one of the airports had to be a category-two airport – this prevented competition on the key routes between main airports. The above example by involving Birmingham (category two) and Brussels (category one) met the condition.

There was also a ceiling on the proportion of passengers that could be in the 5th freedom category – on an annual basis not more than 30% of the people carried could be in the Brussels to Birmingham category. Both the fare directive and the access decision represented minimum degrees of liberalization that had to be accepted by all member states. More flexible arrangements were permitted.

The 'Second Package'

In December 1989 the Council of Transport Ministers returned to the issue of deregulation. A 'Second Package' involving more deregulation was entered into by the Union.

- In respect of the freedom of airlines to make competitive fares, from the beginning of 1993 a system of double disapproval was accepted. Only if both civil aviation authorities on a route refuse to sanction a fare application could an airline be precluded from offering it to its passengers. The regulation essentially provided for a revised system of discount fares based on a reference fare within which all fares meeting the specified criteria were approved automatically. The zones based on the reference, standard fare were 95% to 105% (normal economy), 80% to 94% (discount) and 30% to 79% (deep discount).
- The old system of setting limits to the division of traffic between the bilateral partners was to totally disappear in a phased manner.

- Member States endorsed the principle that governments should not discriminate against airlines provided they meet safety and technical standards and are run economically.
- The Council agreed to address the problem of ownership rules. An airline typically had to be substantially owned by a European state before it could fly from that country. The Council abolished this rule over a two-year period. The implication was that airlines could enjoy 5th freedom rights provided they are registered in the Community.
- Air cargo services were liberalized so that a carrier operating from its home state to another member country can take cargo into a third member state or fly from one member state to another and then to its home state. Cabotage, or operations between two freestanding states, was not liberalized.

The 'Third Package'

The final package of EU air transport reforms came in 1992 to take effect from the following year. This initiated a phased move that, by 1997, resulted in a regulatory framework similar to US domestic aviation although important questions remain regarding the EU's role with respect to international aviation outside the Union.

- The measures removed significant barriers to entry by setting common rules governing safety and financial requirements for new airlines. Charter operators are also allowed to set up in any European market and can sell tickets to private individuals on a seat-only basis at both ends of a service. Since January 1993, EU airlines became able to fly between member states without restriction and within member states (other than their own), subject to some controls on fares and capacity.
- National restrictions on ticket prices were removed with safeguards only if fares fell too low or rose too high.
- Consecutive cabotage was introduced allowing a carrier to add a 'domestic leg' on a flight starting out of its home base to a destination in another member state if the number of passengers on the second leg did not exceed 50% of the total in the main flight. Starting in 1997, full cabotage was permitted, and fares are unregulated.
- Foreign ownership among EU carriers is permitted, and they have, for EU internal purposes, become European airlines. This change does not apply to extra-EU agreements where national bilateral arrangements still pertain. One result has been an increase in cross-share holdings and a rapidly expanding number of alliances.

The adoption of the Third Package essentially moved the EU to a free market situation for airlines within Europe. The discussion of remaining policy reforms within the EEA is, for example, relegated to a few paragraphs in the EU White

Paper on the future of transport policy (Commission of the European Communities, 2001).

Beyond the packages

Recently, the EU Commission has switched its attention to the matter of the relationship between EU air transport policy and external relations (Mencik von Zebinsky, 1996).

The transatlantic market

The traditional right of governments to negotiate bilateral air service agreements with non-EU states has been brought into question by the Commission. This does not mean the EU Commission has not previously been involved in negotiating aviation agreements with third countries. The early agreement between the EU, Norway and Sweden that extended the scope of the EU air transport legislation to the latter countries is an example. The EU Commission was given permission in 1996 through the majority voting procedure to negotiate on behalf of all EU countries on soft issues regarding aviation but did gain the unanimous support for taking over responsibility for hard rights.

The adoption of the offer of Open Skies policies to individual EU member States by the US, and the *de facto* granting of anti-trust immunity to the airlines of countries that follow this course has led to divisions in the EU. In particular, the UK has traditionally opposed giving up negotiating rights with the US, given the domination of Heathrow as the main transatlantic hub. The issue has been complicated by the development of global alliances that fall within the remit of the EU's competition agencies as well as the Transport Directorate.

External matters are an important issue for many carriers that would benefit from opening up larger markets thus offering the opportunity to generate more revenue and contain costs. Within this framework the EU initiated the objective of freeing the North Atlantic air market from economic regulation, a line of argument subsequently refined by the Association of European Airlines (1999) and given support by industrial users of air transport (International Chamber of Commerce, 2000).

This does, however, raise issues of whether external relations are to be left to individual states or are to come under the control of the EU. The Commission effectively tested this by taking a number of EU states before the Court of Justice over their bilateral air service agreements. Technically a judgment was sought on questions of whether many of the extra-EU bilateral air service agreements violated Union laws governing non-discrimination in favor of national suppliers

The European Court of Justice ruling in 2002 has led to significant changes. It effectively said that whilst individual states had responsibility for their own extra-EU affairs, they could not favor their own carriers in bilateral agreements.

In 2003, the Council of Ministers gave over to the Commission the power to negotiate directly with the US to seek a broader and liberalized air market across the North Atlantic. Whether this will lead to major changes is as yet uncertain. If it does, whether the result will be a US-EU Open Skies agreement, or a wider open market entailing such things as common rights of establishment and cabotage is perhaps even less certain. The situation in the Spring of 2004 was the de facto one of stalemate with the US making tentative offers to widen the scope for European investment in US carriers if European airports, including Heathrow, were opened to US carriers. The Europeans were not prepared to accept this.

The enlargement of the European Union

The EU increased in membership by 10 states in 2004 (Cyprus, Czech Republic, Lithuania, Malta, Hungary, Latvia, Poland, Estonia, Slovakia, and Slovenia) bringing its total population to some 450 million. Bulgaria and Romania are scheduled to join in 2007, with Turkey a possible member later. The airlines of the new members are relatively small and in total only generate about $3.9 billion in revenue per annum. (Table AI-1 gives details of the largest operators.) Equally the major airports, while offering extensive regional and charter services, are not large by global standards – e.g., Prague Ruzyne handled 6,315 thousand passengers in 2002; Larnaca, 4,982 thousand passengers, and Warsaw Chopin, 4,937 thousand passengers.

Table AI.1 Major carriers of 2004 EU accession countries (2002 data)

Airline	Revenues ($ millions)	Net result ($ millions)	Passengers (thousands)
LOT Polish Airlines	670	27.7	3,220
CSA Czech Airlines	517	14.6	3,060
Malév Hungarian Airlines	409	-8.9	2,400
Cyprus Airways	306	17.3	1,660
Air Malta	155	-1.1	1660
Adira Airways	106	0.5	810
Lithuanian Airlines	57	9.9	260
Estonian Air	51	2.4	320
AirBaltic	48	0.7	260
Slovak Airlines (estimates)	14	n/a	100

Source: *Airline Business,* February 2004.

The airlines will, however, be confronted with EU rules on state aid, competition policy, and environmental and safety regulations. A number of the markets of the accession countries, most notably Malta and Cyprus have seen competition for many years because of the significant number of charter flights that serve their tourist industries. Some of the carriers already have experience of scheduled competition as domestic markets have opened (e.g., there were three no-frills carriers, including Air Poland and Wizz Air, operating in Poland in 2004 in competition with LOT) but others have been shielded as monopoly national flag carriers. There have also been moves by airlines from the accession countries to reap the benefits of network economies with LOT, for example, joining the Star Alliance and CSA Czech Airlines has become a member of SkyTeam.

The airlines of the 15 Member Union have been active in participating in the markets of the accession countries as opportunities have emerged. This has involved both establishing themselves in larger markets such as Poland and the Czech Republic (British Airways, Air France, Lufthansa, and Austrian Airlines) and in buying stakes in carriers (e.g., SAS Scandinavian Airlines in Estonian Air and AirBaltic).

A major problem likely to be encountered by airlines from the new Member States if they want to expand into the older EU area is gaining slots at the large, congested hub airports. The current EU strategy for slot allocation is unlikely to be helpful to these airlines in this regard. In contrast, there is surplus capacity at the main airports of the accession states (e.g., Bratislava, Budapest, Vilnius, Warsaw, and Riga) that can be exploited by the airlines of the 15-member EU.

European airports policy

Technically, an airport represents a multi-service networked industry with significant monopoly control in the provision of many of its services. Airport infrastructure capacity constraints are crucially important in determining the long-term development of the air transport sector. While the airline industry has been liberalized extensively through the implementation of the EU's Third Package, control over the industry continues to be exercised indirectly or directly by governments through their control of airport capacity allocation and through investment in airport capacity.

Airport pricing policy is of great significance in affecting economically efficient allocations of existing capacity and in signaling where and when expansion of capacity is necessary and justified. It also is a generator of revenues for capacity expansion and up-grading. European airports vary in the economic efficiency but econometric work by Pels (2000) and others also suggests a high level of average inefficiency.

Pricing

Pricing policy, among other things, influences the average size of aircraft at airports, the relative importance and emphasis on short- versus medium- or long-haul services, and the distribution of all EU traffic across the airports system. These factors in turn have important implications for airline network structures – a key competitive tool for carriers in a deregulated market. The initial EU framework for airport charging (Commission of the European Communities, 1995) was slow to be negotiated and met with considerable opposition. In draft form it demanded transparency and non-discrimination in the application of charges to operators. It also allows for willingness-to-pay and Ramsey pricing mechanisms, that, by their nature, are discriminatory.

Ramsey pricing is designed to cover full costs by differential pricing according to a combination of marginal cost considerations and willingness to pay. It adjusts the relative prices for different services according to their costs and the associated elasticities of demand. Further, the directive permits the charges to be related to the airports overall costs or to a regional system of airports' overall costs (Reynolds-Feighan and Feighan, 1997). This 'bundling' of services creates problems when trying to relate charges for particular services to their costs: the pricing signals become obscured.

The airline industry in some parts of Europe has experienced the growth in new products initiated by no-frills scheduled operators, who have been to the fore in driving change in the regions in which they operate. These no-frills carriers have played an important role in bringing about changes to the European air transport sector. Policy-makers explicitly considered the needs of the carriers when planning infrastructure developments. The no-frills carriers have often tended to focus their operations on under-utilized secondary airports close to the key European metropolitan areas. For example, Ryanair, the Irish carrier, has focused its services around Paris (Beauvais), Brussels (Charleroi) and London (Stansted and Luton airports).

It is helpful for policy development to examine the management and traffic distribution in Europe from a system perspective and deal with issues of strategic infrastructure planning from this network point of view. For airports that do not have the ability to expand, because of land constraints or planning regulations, the issue of long term rationing of their fixed capacity and the diversion of traffic should be addressed from this wider viewpoint.

The issue of re-distributing traffic is contentious and critically linked to the role and level of investment funding for airports. The issue needs to be addressed in order for the changes in the economic regulation of airlines to have optimal effect. On the one hand individual services must be costed as precisely as possible, but this is difficult when many items can also be viewed as part of a network or system of facilities. For example, in allocating European airport runway slots, consideration should be given to pavement damage and congestion effects of particular categories of users at individual facilities. These charges then have to be considered in the context of their impacts on other airports in the network, and the possibilities for expansion.

Capacity

There are also legal constraints, partly linked to ICAO pricing principles for international operations, associated with introducing congestion taxes and then using the revenues to provide capacity elsewhere in the system or in the economy more generally. From an investment perspective, while Europe may have sufficient airport infrastructure overall, that capacity is often in the wrong place, with congestion levels growing at the key hub centers. The expansion of capacity is not a straightforward process even if there is land available for further development at existing airports. Environmental regulations and issues related to funding of investment can have a significant impact on the timing and final outcome in the expansion of infrastructure. Air transport has always been a component in multimodal journeys for passengers and freight. Passengers need to get to and from airports, as do freight, so the issue of airport access is an integral part of airport planning and development.

A factor that could prevent enhancement of the interoperability of the aviation sector is the infrastructure constraints and associated delays at airports. Many of the large airports in Europe now have, or have in plan, substantial rail stations with direct links to regional and metropolitan centres. Several airports have sought to integrate high-speed rail interchanges at the airports. Such developments will help to boost air traffic growth. One option is the development of secondary airports in Europe. The developments at this category of airport by no-frills operators has already been discussed.

Economic forces will encourage the larger airports to increasingly substitute long haul services for short haul services, since this will allow for increases in passenger numbers without an accompanying increase in movements. The pattern is already seen at places such as Heathrow. The problem is that all airports rely on the combination of locally originating and transfer passengers to support their air services. So the feasibility of separating out point-to-point traffic and concentrating it at secondary airports is likely to be limited.

Evidence from the US suggests that deregulation stimulated significant growth in air traffic and that US air carriers initially met the increased demand through interactive multiple hub network systems. Point-to-point operators at a certain stage can then enter certain markets where it is possible because of the increased volume to offer direct service. The viability of secondary airports in Europe will depend on the extent of traffic growth, the extent of competition from other surface transport modes and the characteristics of the traffic, particularly the extent of high yield business traffic. These airports will need to offer a certain threshold level of service on routes served since passengers will choose more frequent service (at primary airports) over less frequent service.

Annex II

Airline Deregulation Elsewhere

Introduction

Many countries have been experimenting with market liberalization of their airline markets (Sinha, 2001). We now have considerable global experience with less regulated air transport markets and some of these have been reported in various chapters of the book. Indeed, the movement towards liberalization of the EEA market was stimulated by, amongst other things, the experiences of the 1978 Airline Deregulation Act in the US (Sawers, 1987).

But an important issue is just how far can one expect the experiences of one market to be germane to another. In this case the fact that US airlines have the same immediate financial problems post 2001 may be due to a common generic flaw in the markets in which they operate, or it may be the result of entirely local conditions that coincidentally produce similar outcomes. More likely it is a combination of both with possibly other factors involved.

US domestic deregulation.

Demonstration effects have played their part in forcing regulatory change. The past decade has witnessed considerable liberalization of transport and other markets around the world. Perhaps, one could suggest the stimulus for much of the reform in policy lay in the success of the UK's 1968 Transport Act and its freeing of the road haulage industry from economic regulation. More realistically, however, the essential demonstration effect came from the US, particularly the 1978 Airline Deregulation Act.

The clear initial conclusion of the EU Commission was that European gradualism inherent in the Three Packages was by the mid-1990s leading to a new long-term market equilibrium. The approach was economically superior to the shock therapy adopted under the US 1978 Airline Deregulation Act.

The big bang approach

The view of the European authorities is clear on this. 'The single market in aviation did not occur with a 'Big Bang': there was neither spectacular reduction in fares nor any dramatic disappearance of the more important carriers. Liberalization has happened in a progressive way and without major upsets. This contrasts with the situation that the US experienced at the time of deregulation of the aviation market. The [European] Community has been able to find the correct balance between competition and control mechanisms. Competition and the consumer have both benefited' (Commission of the European Communities, 1996).

Consideration of the respective merits of the gradualist and Big Bang approaches, however, involves a number of different factors. These concern not only the particular features of the favored paths, which others may not be able to replicate, but also the contexts in which the changes were made. The general conclusions of the EU Commission on the outcome of their packages are also open to question.

An argument for sudden and comprehensive regulatory change is that it gives actors in the market, particularly incumbent suppliers, less time to capture the reform process. This advantage can be lost, however, if incumbents play a significant role in defining the format of the Big Bang. Outside of transport this was a demonstrable problem with the privatization of the UK energy sector and most notably the gas industry when the initial privatization gave the industry over to a single supplier essentially on the advice of the former nationalized undertaking.

Conversely, incrementalism provides scope for 'learning-by-doing' and permits policy makers to modify and adjust their reform measures in the light of information gathered as market adjustments take place. Technically, it can also mean that existing hardware can be written down gradually with less physical wastage – there are likely to be fewer stranded costs. But it may also, however, provide the opportunity for those adversely affected to regroup and redefine their reactive positions.

Comparative assessments can be important in the review of policy changes. To be legitimate, however, there is the need to recognize that there can be important differences in background conditions that have to be taken into account. The US reformed its domestic air transport market in a specific way but also in the context of a set of geographical, institutional and economic conditions that were somewhat different from those pertaining to the EU. Further, the implications of changing economic regulatory structures are multidimensional involving complex temporal patterns of effects and a diversity of assessment criteria. While the incrementalism of the EU may have shown advantages in some respects it has been less successful in others.

Because of the size of its air transport industry, and reinforced by its early progress to more liberal markets, the US is often seen as a benchmark in judging developments in other countries. US domestic aviation was heavily regulated from the late 1920s, initially as part of a policy to foster air mail services, but a combination of forces brought about regulatory change in the late 1970s.

A series of academic studies emerged that unfavorably contrasted the performance of regulated interstate aviation with that of some more liberal intrastate services such as those in California and Texas. The fresh intellectual ideas then emerging concerning regulatory capture and market structures brought a new questioning of the underlying rationale for controls. There were also demonstration effects from other network industries, such as UK truckings that liberal markets could be workably competitive and stable. At the macroeconomic level, Keynesian policies seeking to reduce what was perceived as cost push inflationary pressures sought ways of lowering prices.

The 1978 Airline Deregulation Act

A number of de facto measures from 1976 preceded new legislation and the Civil Aeronautics Board (CAB) began permitting discriminatory fare discounts and free access to selected routes. Restrictions on charter services were also relaxed. The subsequent 1978 Airline Deregulation Act was a big bang in the sense that a single act radically changed the way the domestic aviation market was regulated. The move was not to an immediate free market but rather a time schedule for relaxation of price and entry regulations was established such that by January 1983 all fare and entry regulations were eliminated except that carriers must be fit, willing and able. The CAB was abolished in 1985 with its residual functions over such things as international aviation and mergers being transferred to other agencies.

The experience of the US in liberalizing its domestic market demonstrated that regulation had stifled the development of the industry, led to excessive fares, fostered inefficiency and limited consumer choice. Particularly, it impeded the natural growth of 'hub-and-spoke' operations and meant that economics of density and scope could not be fully exploited. In contrast, costs can be considerably reduced in many cases by employing key hubs as consolidation points for flights with routes radiating out from them. Because of the indirect routings and probable changes of planes required at hubs, the replacement of linear direct services by hub-and-spoke operations increases journey times for the passenger, but it also reduces airlines' costs (and in a competitive environment, also fares) while increasing the range of available flight possibilities.

Many of the cost reductions were also a function of the savings brought about by making more efficient use of labor, e.g. in terms of hours flight crew worked. While the conclusions are not unanimous, there is considerable support for the argument that the US liberalization of domestic air transport produced considerable net social benefits in the short to medium term (Morrison and Winston, 1995).

Despite the overall economic gains from deregulation, US reforms have not been trouble-free. Airlines have responded to the new situation by trying to reduce the competition edge of the market.

Because of a concurrent relaxation of anti-trust policy, airline mergers occurred between that limit competition on particular routes and for landing slots at some key hub airports.

The details on computer reservation systems (CRSs) used by travel agents (but owned by airlines) that give customers flight information and seat booking were claimed to have been manipulated to favor the parent company. Even though one of the final acts of the Civil Aeronautics Board in 1985 was to initiate controls over such practices, some 60% of travel agents still wrote tickets on systems provided by United or American. Bonus schemes and other incentives also suggest that the resulting 'halo effects' bias the system, even if the information held within the CRS program is objectively presented. Frequent-flyer programs giving extra flights to loyal customers have been deployed as a defensive weapon to encourage

Wings Across Europe

existing passengers to stay with the airline currently favored but can also act as an offensive device to attract flyers away from companies not offering such a perk.

Despite these difficulties, the general view seems to be that in the decade following deregulation that the changes yielded a positive rate of return to the air transport industry as a whole. While market imperfections exist and the post-reform market hardly corresponds to the ideals established as benchmarks for efficiency by traditional neo-classical economists, distortions are seen as less damaging than the government failures which accompanied the previous, heavily regulated regime. Policy makers in Europe have seen this as a case for liberalization of their markets, albeit in a modified form, designed to contain the major problems that have arisen in the US.

More recently the situation in the US has become more opaque. The major network carriers are suffering from severe financial problems that are only partly related to the events of September 2001, and the subsequent SARS outbreak and Gulf conflict, and are persisting in many cases (Table AII-1). A number of large carriers, most notably US Airways and United Airlines have gone into Chapter 11 bankruptcy and by 2004, Delta was speculating on the possibility of it following suit in 2005.

Table AII.1 Profits of US carriers in 2002

Airline	Profit (Loss)	Margins
Jetblue	$105	16.5%
Southwest	$ 417	7.6%
Airtran	$31	4.2%
Continental	$(312)	-3.7%
Alaska	$(75)	-4.1%
Northwest	$(846)	-8.9%
Delta	$(1,309)	-9.8%
America West	$(203)	-10.1%
US Air	$(1,200)	-18.2%
American	$(3,330)	-19.2%
United	$(2,837)	-19.9%

Note: Figures in parenthesis indicate loses.

The impact of low cost carriers, most notably Southwest and JetBlue, has been to force down fares in many markets. There had been significant increases in labor costs particularly following the pay rise awarded to United Airline pilots in 2000, and these have proved to be difficult to claw back in a depressed market. Demand patterns also seem to be shifting. The extensive use of air travel by the high technology sector in the late 1990s, with a boom in demand for business class fares has been tempered by the more recent economic weakness of the sector and the slowing of the US economy more generally from 2000.

The US and European air transport markets

The debates surrounding the need to reform European aviation in the 1980s and 1990s were very clearly influenced by the deregulation of the US domestic air transport market, as well as what was going on in a variety of other, not always transport, markets. The reforms in EU air transport policy came about at a time when the results of the first ten years of US deregulation were beginning to emerge.

Experiences observed in one market may, however, have limited relevance for what might occur elsewhere, so it is important to examine ways in which European air transport deviates from the US domestic sector. How much weight one can attach to the US experience is thus judgmental but inevitably must take into account a number of factors. As can be seen from a brief listing of the main differences (Table AII.2) these extend across a wide range of both demand and supply considerations as well as institutional factors.

Air transport reforms elsewhere

While the focus has been on comparisons between the US and European aviation, policy changes elsewhere should not be ignored. These other changes, while attracting less attention, have interesting features of their own and offer insights into the policy reform process. Space precludes a detailed discussion and an in-depth account would be moving away from the main thrust of the book. Therefore, only some brief comments are offered.

Canada

The Canadian aviation situation has historically had some features similar to those found in Europe. One of its major carriers, Air Canadia, was traditionally state owned and heavily protected in terms of both fares and market access. The path of reform has also been similar to the EU in that it has been incremental. There are, however, also important differences. The Canadian market is essentially linear, east west. The scope for hub-and-spoke operations is thus limited. There are vast areas in northern Canada that are only effectively accessible by air, and social and political cohesion requires that adequate services be offered in such areas.

The initial changes began in 1979 with removal of all access controls in the long distance Canadian market and the allowing of CP Air to compete with Air Canada. It also introduced the charter carrier Wardair into the scheduled market. Limited fare discounting was permitted. A series of other minor changes led in 1987 to the National Transport Act that, in effect, cut the Canadian domestic market in two. The northern part remained protected with some directly subsidized services while there was effective US style deregulation in the southern part.

Table AII.2 Difference between the US and European air transport markets

- *Domestic/international traffic split.* The vast majority of EU carriage and historically its airlines have largely operated in regulated international markets, and the natural inclination for rent protecting may take longer to erode than occurred in the US.
- *The non-scheduled market.* European aviation involves a large charter component. In 1999, the number of RPKs done by non-scheduled services in the US was 12 billion out of a total of 772 billion – comparable figures for Europe were 120 billion out of 337 billion.
- *Market size.* The average route length in Europe is 720 kilometers; in the US it is 1220 kilometers. The average number of passengers per scheduled route is about 100,000 in Europe but virtually double that in the US.
- *Airline size.* The scale of the European market is also reflected in the actual size of the European Union's airlines. The merger of British Airways (46.3 billion scheduled passenger-kilometers) and British Caledonian (8.8 billion) in 1987 made it the largest European carrier in terms of passenger-kilometers. The situation has not changed significantly since that time and while there are some large European carriers, the world's largest passenger airlines and cargo carriers as measured by conventional parameters are US based.
- *Ownership of airlines.* While the US commercial airline industry has always been in private hands; most major European carriers have been state owned and still a number of carriers remain state owned or have a significant state involvement.
- *Approaches to bankruptcy.* Chapter 11 arrangements in the US are less stringent than most European bankruptcy regimes and allow for existing management, under supervision, to restructure a company's finances rather than have the assets of the undertaking realized.
- *Subsidies.* The US had no tradition of explicitly subsidizing airlines until 2001, except in limited cases social air services. There has been a tradition of subsidizing European flag carriers but EU policy on state subsidies has now been tightened and criteria made explicit under which subsidies may be given.
- *Intermodal competition.* There is substantial intermodal competition in Europe from high-speed train services.
- *Infrastructure availability.* US air traffic control is centralized, whereas air traffic control in Europe has been a national concern. The EU system comprises a patchwork, while the US has less than half the number of centers and standardized mainframe computers. Airport capacity in Europe is rapidly being reached at many major terminals whereas this is a more limited problem in the US. More than 50% of traffic in Europe passes through 24 airports and, although figures are somewhat subjective on the issue of airport capacity, all of these have reached their technical capacity or are very close to it.
- *Advantage of hindsight.* EU policy makers and airlines have the US as an example to warn them against some of the difficulties to be avoided.

The state owned Air Canada was committed to act commercially and was subsequently privatized. There was a considerable amount of merger activity among smaller, regional carriers and acquisitions in the Canadian market after deregulation but Air Canada and Canadian International effectively became a duopoly over the major routes. In 1996 the Canadian and US governments agreed on a phased relaxation of their bilateral agreement that after three years would effectively integrate their air transport markets. US carriers integrated their activities with Canadian airlines through equity holdings and code sharing.

The performance of Canadian International deteriorated in the late 1990s and Air Canada subsequently absorbed the company and assumed a closely monitored quasi monopoly position in the market. The overall result of the Canadian reforms have been increased efficiency in terms of airline costs although, after an initial phase of market entry, mainly by regional carriers into the national market, there has been considerable rationalization and financial restructuring. In 2002, Air Canada began to make a small profit, but by the end of 2003 had sunk to making record losses.

Reforms of the airline regulator structure have been accompanied by both the corporatization of the air traffic control system, and the removal of most airports from government control; airports are now largely the responsibility of provincial or local authorities. The aims of these moves have been very much to enhance efficiency and to force management to view the provision of air traffic control and airport infrastructure in more commercial terms.

Australia and New Zealand

In the early 1980s, both Australia and New Zealand had separate, tightly regulated airline markets. In spite of the remoteness of the two countries and the close economic ties between them, each had a distinct domestic aviation market and, in the case of Australia, there was no integration of the international and domestic markets, which were served by different airlines. Domestically, the Australian market was served by two airlines (the 'Two Airline Policy'); Ansett, which was privately owned, and the state owned Qantas that also enjoyed a monopoly in terms of international services. The state owned Air New Zealand enjoyed market powers in both its domestic and international markets.

By the mid-1990s, the aviation markets of both countries have undergone extensive change. Domestic markets have been deregulated, and there is scope for open competition, although the actual number of competitors is not large. New airlines have attempted entry in both markets but there has been considerable numbers of failures. There was some progress towards forming a single aviation market covering both countries, although this was subsequently put on hold. At the same time, there has been considerable change at the corporate level. The competitive situation in Australia became so acute in 2001 that Ansett went into bankruptcy and left a vacuum for the expansion of the established incumbent, Quantas, and the relatively new entrant, Virgin Blue.

Comparative assessments of public policy reform are often very insightful. To be so, however, there is the need to recognize that there can often be important differences in background conditions that have to be taken into account. When the US reformed its domestic air transport market it did so in a specific way. But it also did it in the context of its particular geographical, institutional and social conditions that were somewhat different to those pertaining to the EU at the same time. The same can be said of countries such as Canada, Australia and New Zealand that have also been the subject of recent deregulation studies.

Glossary of Terms

AEA. Association of European Airlines.

Airline alliances. These are agreements between carriers to coordinate their activities in various ways. Such coordination may include code sharing, reciprocal frequent flier programs, common lounge facilities, and coordinated schedules. In some cases these agreements may relate to single point-to-point services but in other cases they may be more wide ranging. Major strategic alliances involve several carriers and are intercontinental in their coverage.

Air Service agreements. Inter government agreements that establish the conduct of trade in international air services. Traditionally these have been mainly negotiated between two countries but there has recently been a move towards more multilateral agreements. The provisions of agreements vary but can include market access, route freedoms granted, capacity, flight frequency, and methods of tariff determinations.

ATA. Air Transport Association of America.

Barriers to entry and/or exit. Legal, institutional and economic factors that limit the ability of (both new and existing) airlines to enter or exit a particular market.

Cabotage. The right of foreign-owned airlines to provide commercial domestic air services in a host country.

CAB. Civil Aeronautics Board.

Charter operations. Air services that do not operate to a regular schedule.

Chicago Convention. International meeting in 1944 that established the framework for international agreements on air transport services.

Code-sharing. A marketing arrangement between airlines allowing them to sell seats on each other's flights under their own designator code. In the case of connecting flights of two or more code-sharing carriers, the whole flight is displayed as a single carrier service on a CRS.

Computer Reservation System (CRS). The electronic data management system that distributes information, availability status and the price of travel services to retailers and directly to consumers.

Consolidation. Reduction in the number of airlines serving a market either through airline closure – allowing remaining airlines an opportunity to increase their market share – or through the acquisition of an airline by another airline. Airline consolidation results in increased market concentration among airlines remaining in the market.

Contestability. An economic theory that emphasizes the importance of potential entry by new firms for efficient price and service outcomes. The more effective the threat of entry, the smaller any excess profit accruing to existing airlines servicing the market.

Economies of scale. Average unit cost of production declines as airline output increases.

Economies of scope. One airline can produce two or more services more cheaply than if those same services were produced separately by different carriers.

Economies of traffic density. Average unit cost of production declines as the amount of traffic increases between any given set of points served.

Efficiency. Economic measure of the best use of resources.
- Allocative efficiency refers to the optimum allocation of scarce resources between end-users in order to produce a combination of goods and services that best meets the pattern of consumer demand.
- X-efficiency refers to the effectiveness of an airline's management in minimizing the costs of producing a given level of output.

Elasticty. The proportional responsiveness of the quantity demanded to changes in factors such as income and price.

Empty core. A market condition whereby it is not possible for suppliers to recover their full costs.

ECAC. European Civil Aviation Conference.

EEA. The European Economic Area that embraces the European Union countries together with Iceland and Norway but excludes Gibraltar and the Azores.

EU. European Union.

Flag carrier. Countries with only a government-owned airline often identify the airline as the national flag carrier.

Freedoms of the Skies. International aviation rights of passage:

- 1st freedom. The right of an airline of one country to fly over the territory of another country without landing.
- 2nd freedom. The right of an airline of one country to land in another country for non-traffic reasons, such as maintenance or refueling, while en route to another country.
- 3rd freedom. The right of an airline of one country to carry traffic from its country of registry to another country.
- 4th freedom. The right of an airline of one country to carry traffic from another country to its own country of registry.
- 5th freedom. The right of an airline of one country to carry traffic between two countries outside its own country of registry as long as the flight originates or terminates in its own country of registry.
- 6th freedom. The right of an airline of one country to carry traffic between two foreign countries via its own country of registry. This is a combination of the third and fourth freedoms.
- 7th freedom. The right of an airline to operate stand-alone services entirely outside the territory of its home state, to carry traffic between two foreign states.
- 8th freedom. The right of an airline to carry traffic between two points within the territory of a foreign state (cabotage).

Frequent flyer programs. Schemes offering flight or other benefits to travelers who fly frequently with an airline. These programs are expanding to include services unrelated to air travel.

Full service carriers. These are network carriers that provide a full range of hub-and-spoke services (often also operating point to point services as well) and offer a range of service quality options (e.g., business class and economy class).

Hub-and-spoke networks. An airline operating structure where one or more airports act as a focus for an airline's operations. In a hub and spoke network, traffic is collected from a number of 'spoke' or feeder points and consolidated at the hub point prior to redistributing traffic out of the hub to connect with flights to another destination.

IATA. International Air Transport Association.

ICAO. International Civil Aviation Organization.

Interlining. Carriage of passengers and freight by one airline on behalf of another airline, based on a formal arrangement (an interline agreement) between the airlines. Carriers involved in such an interlining agreement are required to honor tickets issued by other carriers in the agreement. The identity of each carrier is maintained.

Market distortions.
- **Market failure**. The market may fail to produce an efficient outcome because of intrinsic features such as the existence of natural market power or high fixed costs.
- **Government intervention failure**. Intervention by government in markets (e.g. through regulations) may lead to a lessening of economic efficiency.

Market power. The ability of a supplier to influence the market price for the product being sold. It normally entails a degree of market power.

Monopoly. Supply by a single seller.

No-frills carriers (low cost carriers in the US). These are carriers that provide a basic point to point service, often between secondary airports.

Non-scheduled (or charter) services. Flights performed for remuneration on an irregular basis.

OECD. Organisation for Economic Cooperation and Development.

Open Skies Agreements. Liberal air service agreements initiated by the US. Initially they varied in their form but subsequently conform to a standard template. A combination of bilateral Open Skies Agreements can form the basis for a liberal multilateral air transport market.

Reservation System. A net of computer terminals through wich a travel agent researches air travel arrangements, and issues tickets.

Scheduled services. Flights listed in a published timetable, or so regular and frequent as to constitute a recognizably systematic series, and performed for remuneration.

Slots. The right to land/take-off from an airport at a specified time.

Structural adjustment. Productive undertakings alter their methods of and approaches to production or the nature of their output in response to changes in consumer demands, technological shifts or developments in the competitive framework.

Tendering. A bidding system whereby the right to supply a non-profit service is awarded to the supplier requiring the lowest subsidy.

Value chain. The series of links in the production process that add value in the provision of the final good or service.

Workable competition. A situation where competition is not perfect but where any additional government intervention in the market will not add to efficiency.

WTO. World Trade Organisation.

X-inefficiency. See **Efficiency.**

Yield management. This is a dynamic form of price discrimination whereby rather than an identical fare being charged to each passenger, users are charged different fares reflecting their willingness to pay.

References

Alamdari, F. (1997) Airline labour cost in a liberalised Europe, paper to the International Air Transport Conference, Vancouver.

Aschauer, D.A. (1989) Why is infrastructure important?, *Journal of Monetary Economics*, 23: 177-200.

Association of European Airlines (1997a) *User Costs at Airports in Europe, Asia and the USA*, AEA, Brussels.

Association of European Airlines (1997b) *Pilots Flight Duty and Rest Times*, AEA, Brussels.

Association of European Airlines (1999) *Towards a Transatlantic Common Aviation Area: AEA Policy Statement*, AEA, Brussels.

Association of European Airlines (various years) *Yearbook*, AEA, Brussels.

BAE Systems (2000) *Updating and Development of Economic and Fares Data Regarding the European Air Travel Industry: 2000 Annual Report*, BAE Systems, Chorley.

Barkin, T.I., Hertzell, O.S, and Young, S.J. (1995) Facing low-cost competition: lessons from US airlines, *The McKinsley Quarterly*, 4: 87-99.

Barrett, S.A. (1987) *Flying High: Airline Prices and European Regulation*, Avebury, Aldershot.

Barrett, S.A. (1990) Deregulating European aviation – a case study, *Transportation*, 16: 311-327.

Baumol, W.J., Panzar, J.C. and Willig, R.D. (1982) *Contestable Markets and the Theory of Industrial Structure*, Harcourt Brace Jovanovich, New York.

Binggeli, U. and Pompoe, L. (2002) Hyped hopes for Europe's low cost airlines, *McKinsey Quarterly*, October: 1-8.

Bittlingmayer, G (1982) Decreasing average cost and competition, *Journal of Law and Economics*, 25: 201-209.

Bittlingmayer, G. (1985) Did antitrust policy cause the great merger wave?, *Journal of Law and Economics*, 28: 77-118.

Boeing Commercial Airplanes (2002a) *Current Market Outlook*, Boeing, Seattle.

Boeing Commercial Airplanes (2002b) *World Cargo Forecast*, Boeing, Seattle.

Breyer, S. (1984) *Regulation and its Reform*, Harvard University Press, Cambridge.

Brooks, M. and Button, K.J. (1995) Separating track from operations: a look at international experiences, *International Journal of Transport Economics*, 22: 235-260.

Brueckner, J.J. (2002) Airline traffic and urban economic development (mimeo).

Brueckner, J.J. (2003) The benefits of code sharing and antitrust immunity for international passengers, with an application to the Star Alliance, *Journal of Air Transport Management*, 9: 83-91.

Brueckner, J.J. and Whelen, W.T. (2000) The price effect of international airline alliances, *Journal of Law and Economics*, 43: 503-545.

Button, K.J. (1996) Liberalising European aviation: is there an empty core problem, *Journal of Transport Economics and Policy*, 30: 275-291.

Button, K.J. (1998) Infrastructure investment endogenous growth and economic convergence, *Annals of Regional Science*, 32: 145-163.

Button, K.J. and Taylor, S.Y. (2000) International air transportation and economic development, *Journal of Air Transport Management*, 6: 209–222.

Button, K.J., Haynes, K. and Stough, R. (1998) *Flying into the Future: Air Transport Policy in the European Union*, Edward Elgar, Cheltenham.

Button, K.J., Lall, S., Stough, R. and Trice, M. (1999) High-technology employment and hub airports, *Journal of Air Transport Management*, 5: 53-59.

Button, K.J., Costa, A. and Reis v. (2004) How to control a route from the supply side: the case of TAP - Air Portugal, paper to the Air Transport Research Society Annual World Conference, Istanbul.

Caves, D.W., Christensen, L.R., Tretheway, M.W. and Windle, R.J. (1987) An assessment of the efficiency effects of U.S. airline deregulation via an international comparison, in Bailey, E.E. (ed.) *Public Regulation: New Perspectives on Institutions and Policies*, MIT Press, Cambridge.

Coughlin, C.C., Cohen, J.P. and Khan, S.R. (2002) Aviation security and terrorism: a review of the economic issues, *Federal Bank of St Louis Review*, September/October: 9-24.

Comité des Sages for Air Transport (1994) *Expanding Horizons*, European Commission, Brussels.

Commission of the European Communities (1995) *Consultation paper on Airport Charges, Commission of The European Communities*, Directorate General VII, CEC, Brussels.

Commission of the European Communities (1996) *Impact of the Third Package of Air Transport Liberalization Measures*, COM (1996) 415 Final, Brussels.

Commission of the European Communities (1999) *Communication of the Commission of the Council and the European Parliament on the Creation of a Single European Sky*, COM (1999) 614 Final, Brussels.

Commission of the European Communities (2001) *Regulation of the European Parliament and of the Council on Establishing Common Rules on Compensation and Assistance to Air Passengers in the Event of Denied Boarding and of Cancellation or Long Delay of Flights*, COM (2001) 784, Brussels.

Commission of the European Communities (2002) *Regulation of the European Parliament and of the Council on Insurance Requirements for Air carriers and Aircraft Operators*, COM (2002) 521, Brussels.

Coy, P. (2002) Deregulation: innovation vs. stability, *Business Week*, January 28: 108-109.

Cranfield University Air Transport Group (2002) *A Study of Low Cost Carriers and Network Quality*, Cranfield University.

Distexhe, V. and Perelman, S. (1994) Technical efficiency and productivity growth in an era of deregulation: the case of airlines, *Swiss Journal of Economics and Statistics*, 130: 669-689.

Doganis, R. and Dennis, N.P.S. (1989) Lessons in hubbing, *Airline Business*, March: 42-45.

Dresner, M. (2004) Many fields of battle: how cost structure affects competition across multiple markets, paper to the Air Transport Research Society Annual World Conference, Istanbul.

Dresner, M., Flipcop, S. and Windle, R. (1995) Trans-Atlantic airline alliances: a preliminary evaluation, *Journal of the Transportation Research Forum*, 35: 13-25.

Encanoua, D. (1991) Liberalizing European airlines: cost and factor productivity evidence, *International Journal of Industrial Organization*, 9: 109-124.

European Cockpit Association Industrial Sub-group (2002) *Low Cost carriers in the European Aviation Single Market*, ECA, Brussels.

Francis, G., Alessandro, F. and Humphreys, I. (2003) Airport-airline interaction: the impact of low-cost carriers on two European airports, *Journal of Air Transport Management*, 9: 267-273.

Gellman Research Associates (1994) *A Study of International Airline Code Sharing*, Office of Aviation and International Economics, Office of the Secretary of US Department of Transportation, Washington.

Good, D.H., Röller, L-H. and Sickles, R.C. (1993) US airline deregulation: implications for European transport, *Economic Journal*, 103: 1028-1041.

Good, D.H., Röller, L-H. and Sickles, R.C. (1995) Airline efficiency differences between Europe and the US: implications for the pace of EC integration and domestic regulation, *European Journal of Operations Research*, 80: 508-518.

Hansson, T., Ringbeck, J. and Franke, M. (2002) *Airlines: A New Operating Model. Providing service and Coverage without the Cost Penalty*, Booz, Allen, Hamilton, Los Angeles.

Horan, H. (2002) What is the future of the European flag-carrier?, *Aviation Strategy*, 59: 2-15.

ICF Consulting and Button, K.J. (2003) *Overview of Europe's Aviation Industry: Structure and Competition*, ICF, Fairfax.

International Chamber of Commerce (2000) *Policy Statement on the Need for greater Liberalization of International Air Transport*, ICC, Paris.

Juan, E.J. (1995) *Airport Infrastructure: The Emerging Role of the Private Sector*, CFS Discussion Paper Number 115, World Bank, Washington.

Kahn, A.E. (1988) *The Economics of Regulation: Principles and Institutions*, MIT Press, Cambridge.

Kahn, A.E. (2001) *Whom the Gods Would Destroy, or How not to Deregulate*, American Enterprise Institute/Brookings, Institution, Washington.

Mandel, B.N. (1999) *Airport Choice and Competition: A Strategic Approach*, Mkmetric GmbH, Karlsruhe.

Mason, K., Whelen, C. and Williams, G. (2000) *Europe's Low Cost airlines: An Analysis of the Economics and Operating Characteristics of Europe's Charter and Low Cost Scheduled Airlines*, Air Transport Group Research Report 7, Cranfield University.

McGowan, F. and Seabright, P. (1989) Deregulating European airlines, *Economic Policy*, October: 283-344.

Mencik von Zebinsky, A.A. (1996) *European Union External Competence and External Relations in Air Transport*, Kluwer, The Hague.

Mercer Management Consulting (2002) *Developments in the Airline Industry – Impact of Low Cost Carriers*, Mercer Management Consulting, Munich.

Merrill Lynch (2002) *European Transport: In Search of Long Term Value*, Merrill Lynch Global Securities Research and Economic Group, London.

Morrison, S.A. and Winston, C. (1995) *The Evolution of the Airline Industry*, Brookings Institution, Washington.

Ng, C.K. and Seabright, P. (2001) Competition, privatization and productive efficiency: evidence from the airline industry, *Economic Journal*, 111: 591-619.

Oum, T.H. and Yu, C. (1995) A productivity comparison of the world's major airlines, *Journal of Air Transport Management*, 2: 181-195.

Papatheodorou, A. (2002) Civil aviation regimes and leisure tourism in Europe, *Journal of Air Transport Management* (forthcoming).

Parkinson, J. and Sentance, A. (2002) *Airport capacity and the future of European hub airports*, to the 6th Air Transport Research Society Conference, Seattle.

Pels, E. (2000) *Airport Economics and Policy: Efficiency, Competition, and Interaction*, PhD. Thesis, free University, Amsterdam.

Pirrong, S.C. (1992) An application of core theory to the analysis of ocean shipping markets, *Journal of Law and Economics*, 35: 89–131.

Pitfield, D. (2004) A time series analysis of the pricing behaviour of directly competitive 'low-cost airlines', paper to the 5th World Congress of the Regional Science Association International, Port Elizabeth.

Porter, M.E. (1985) *Competitive Advantage: Creating and Sustaining Superior Performance*, Free Press, New York.

Reynolds-Feighan, A.J. and Feighan, K.J. (1997) Airport services and airport charging systems: a critical review of the EU Common Framework, *Transportation Research E*, 33: 311-320.

Rietveld, P. (1997) *Drie Mianpoortssystemen voor Nederland*, Economische Statistische Berichten, The Hague.

Sawers, D. (1987) *Competition in the Air – What Europe Can Learn from the USA*, Institute of Economic Affairs Research Monograph 41, London.

Schipper, Y. (1999) *Market Structure and Environmental Costs in Aviation: A Welfare Analysis of European Air Transport Reform*, PhD dissertation, Free University of Amsterdam.

Science Applications International, MKmetric GmbH and Institut für Weltwirtscahft (2000) *The Impact of Liberalizing International Aviation Bilaterals on the Northern German Region*, Hamburg.

Sinha, D. (2001) *Deregulation and Liberalization of the Airline Industry*, Ashgate, Aldershot.

Sjostrom, W. (1989) Collusion in ocean shipping: a test of monopoly and empty core models, *Journal of Political Economy*, 97: 1160-1179.

Sjostrom, W. (1993) Antitrust immunity for shipping conferences: an empty core approach, *Antitrust Bulletin*, 38: 419-423.

Smith T.K. (1995) Why air travel doesn't work, *Fortune*, April 3: 42-56.

Starkie, D., and Starrs, M. (1984) Contestability and sustainability in regional airline markets, *Economic Record*, 60: 274-283.

Suen, W.A. (2002) Alliance strategy and the fall of Swissair, *Journal of Air Transport Management* (forthcoming).

Telser, L.G. (1978) *Economic Theory and the Core*, University of Chicago Press, Chicago.

Telser, L.G. (1985) Cooperation, competition, and efficiency, *Journal of Law and Economics*, 28: 271-295.

Transportation Research Board (2002) *Freight Capacity for the 21st Century*, TRB, Washington.

UK Civil Aviation Authority (1988) *Statement of Policies on Air Transport Licensing – June 1988*, CAP539, CAA, London.

UK Civil Aviation Authority (1993) *Airline Competition in the Single European Market*, CAP623, CAA, London.

UK Civil Aviation Authority (1994a) *The Economic Impact of New Air services: A Study of Long Haul Services at UK Regional Airports*, CAP 638, CAA, London.

UK Civil Aviation Authority (1994b) *Airline Competition on European Long Haul Routes*, CAA, London.

UK Civil Aviation Authority (1998) *The Single European Aviation Market: The First Five Years*, CAA: London.

UK Office of Fare Trading (1999) *Assessment of Individual Agreements and Conduct*, OFT, London.

Upham, P., Maughan, J., Raper, D. and Thomas, C. (eds) (2003) *Towards Sustainable Aviation*, Earthscan, London.

US General Accounting Office (1995) *International Aviation: Airline Alliances Produce Benefits but Effect on Competition is Uncertain*, GAO/RCED-95-99, Washington.

US National Commission to Ensure a Strong Competitive Airline Industry (1993) *Change, Challenge and Competition*, US Government Printing Office, Washington.

US Transportation Research Board (1999) *Entry and Competition in the US Airline Industry: Issue and Opportunities*, TRB, Washington.

Williams, G. (2001) Will Europe's charter carriers be replaced by 'no-frills' scheduled airlines, *Journal of Air Transport Management*, 7: 277-286.

Zonneveld, G. (2002) *European Airlines Review: Low Cost Carriers*, WestLB Panmure, New York.

Index